Flowers of the Southwest Mountains

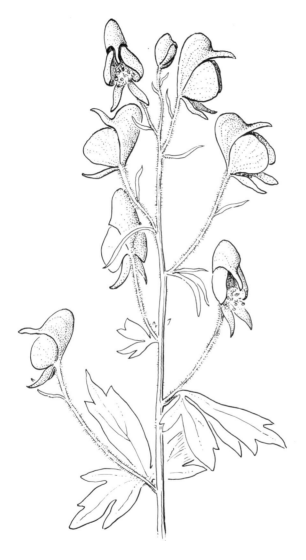

written by
Leslie P. Arnberger

drawings by
Jeanne R. Janish

Editorial: T.J. Priehs, Carolyn Dodson
Design: Christina Watkins
Production: T.J. Priehs, Christina Watkins,
Nancy Curtis, Bob Petersen
Lithography: Lorraine Press

Southwest Parks and Monuments Association
wishes to express its sincere thanks to Dr. Charles
Mason and Mr. George Yatskievych of the University
of Arizona for reviewing the manuscript.

221 North Court
Tucson, Arizona 85701

Contents

COVER: Columbia monkshood, *Aconitum columbianum*

Acknowledgments

Sincere thanks are due the late Dr. T. H. Kearney of the California Academy of Sciences for his helpful suggestions in selecting the plants to be described in this book and for checking the manuscript for botanical accuracy.

To the late Robert H. Peebles, also, the author is sincerely grateful for his generous help in checking the list of plants.

For the privilege of borrowing herbarium specimens used in the preparation of drawings, both author and illustrator offer their genuine appreciation to Chester F. Deaver, formerly of Northern Arizona University, Flagstaff, Arizona, Dr. Howard Dittmer of the University of New Mexico at Albuquerque, and Roxana Stinchfield Ferris, Assistant Curator of the Dudley Herbarium of Stanford University. Last, but certainly not least, my thanks to Jeanne R. Janish, an exceedingly talented illustrator, whose accurate and attractive drawings contribute so much to this book.

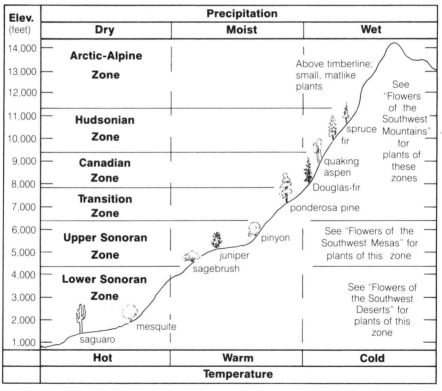

Idealized profile of the Southwest from desert to mountaintop showing characteristic plants of the different life zones.

Introduction

Flowers and plants, like people, are much more interesting when we know their names and a little bit about them. This booklet has been written to introduce you to some of the more common and beautiful plants of the Southwest mountains, and it should prove a helpful companion on your trips afield. In addition to giving the plant names, it provides interesting information on uses of plants by the native peoples and by animals. Every effort has been made to keep written botanical description to a minimum, with the thought that the excellent drawings are the most important part of the book, and will be of the greatest help in identification. Very few technical terms have been used, so if you don't know a peduncle from a petiole, don't worry about it. It is not necessary.

This booklet is specifically designed as a guide to the mountain flowers of Arizona, New Mexico, Colorado, and Utah. Of course, not all of the species described are found in all four states, but the great majority of them are. In fact, some of the plants are found in the higher and colder places of the entire west, and a few are so widespread that they grow from coast to coast. For our purposes we shall consider the mountains to start at an elevation of about 7,000 feet. At this altitude we are in what is called the Transition Life Zone, and the most conspicuous plant is the large ponderosa pine.

To be included in this book, it was necessary for a plant to be either common or conspicuous, and preferably both. It is impossible to choose 150 plants that are equally abundant throughout the Southwest, so cases may be found where one species is common in Arizona and New Mexico, rare in Colorado and does not even grow in Utah. However, you can tell at a glance if the plant grows in the state where you are, by looking at the statement on distribution given for each species.

This booklet is the last of a triad which includes two others titled "Flowers of the Southwest Deserts" and "Flowers of the Southwest Mesas." The "Desert Book" introduces flowers of the hot, dry deserts of Arizona, New Mexico, Texas, and California while the "Mesa Book" covers the common plants of the pinyon-juniper woodland extending from about 4,500 feet to 7,500 feet in elevation. From there on up, to away above timberline, this "Mountain Book" is the one to use.

It is unfortunate that most botanists write only for other botanists, so that the great majority of us who are interested in wildflowers find it exceedingly difficult to recognize our flowered friends through the use of standard botanical books. It is with this thought in mind that I, a layman, have undertaken the preparation of this book. To make it useable for the amateur I have been forced to take shortcuts, disregard several established botanical principles, and convert technical terms into understandable every-day language. For those offenses the botanists have my sincere apologies. We must all appreciate, however, the work of many botanists over many years, for without their detailed knowledge this book and others like it could never have been written.

How to use this book

Several methods have been developed for identifying plants, but, unfortunately, most of those keys are much too involved for the amateur to use. In standard botanical books the plants are arranged in order of their relationship to one another; that is, primitive plants such as ferns are first, and the more highly developed plants such as sunflowers are last. The keys used in such books, although quite accurate, are technical, and the ordinary amateur usually gives up in total confusion long before he has identified the flower he is interested in.

To aid in identification the plants described in this booklet have been grouped according to plant form and flower color into seven sections as follows:

(1) Trees — includes tree size plants not having conspicuous flowers. (pp. 9-16)
(2) Whitish — White, cream, greenish-white. (pp. 21-48)
(3) Red. (pp. 49-52)
(4) Pinkish—lighter shades of red to whitish, tinged with pink.(pp.53-64)
(5) Yellowish — yellow to orange. (pp. 69-96)
(6) Bluish — shades of blue, violet, and sometimes purple. (pp. 101-116)
(7) Purplish — purple, lilac, rose- and reddish-purple. (pp. 117-128)

With the exception of the section on trees this book uses a key based upon the color of the flower. This key is by no means as accurate as the regular botanical key, but with practice the amateur should be able to make a general identification, at least to the genus.

The greatest weakness in a color key is the variability of flower colors within the same species. For this reason the color groups have not been broken down into the various shades of white, yellow, red, etc. Instead, I have tried to lump all related colors as much as possible, as in the case of whitish, which includes cream-colored and greenish-white flowers as well as white. There are many examples of color variation in the same species such as the sidebells pyrola which has flowers that vary from white to greenish-white, or the flowers of the silvery lupine which may range in color from blue to purple. A more complete description of the flower color for each species is given in the written description.

Let's say that you are walking along a mountain trail, and come upon an interesting looking plant to which you would like an introduction. The plant has many small, yellow flowers clustered along the top of the stem; it is about 5 feet high, and has large leaves which are so thickly covered with hair that they feel furry.

The first thing to do is to decide upon the color of the flowers. In this case they are yellow, so turn to the yellow section of the book, and slowly leaf through the pages looking for a drawing of a plant that resembles yours in size, shape of leaf, and location of flowers. Without too much difficulty you should decide that the drawing of the common mullein looks the most like yours. Now that you have gotten that far, check the text to see that the plant grows where you are, blossoms at the right time of year and, in the case of the mullein, you will find a statement to the effect that the leaves are covered with dense hair. If all of the details in drawing and text check with your plant, chances are that you have correctly identified your plant or at least one of its close relatives.

As another example, suppose you find a tree with which you are not acquainted. It is a large tree and has needles and bears cones with odd looking three-pointed bracts sticking out among the cone scales. As this plant is a tree, you would turn to the tree section in the front of the book. First look at the identification chart for evergreens on page 8, and look at the drawings for a tree that has the same kind of leaves (needles) and cones that yours has. You can decide right away that your tree is not a pine, as the needles are borne singly instead of several to a bundle, and the difference in cones will tell you it is not a spruce or fir. The only thing left is the Douglas-fir, and this is the tree you are looking for, as the drawing of the cone shows the same three-pointed bracts that you noticed on the cones on the tree. You can double-check this identification now by turning to page 12 where the Douglas-fir is pictured and described in detail.

The two examples given above are easy ones, chosen to illustrate the basic method of using this booklet. It will not be that easy in the case of many flowers which you may attempt to track down.

Unfortunately, the plants have not read the books, and so they persist in being somewhat different from the way they are described. For instance, the beautiful blue columbine may be white under some circumstances and the usually towering Engelmann spruce may be nothing more than a straggling shrub at timberline. One plant, the common juniper, has been a problem child, for it refused to fit under the section on trees, as it is a low, prostrate shrub, and, because it has no flowers to speak of, it could not be included under one of the sections on color. Finally, out of desperation, it was included in the tree section because its close relatives, to which it bears a resemblance, are all trees.

Don't be disappointed if you fail to identify all the flowers you find, for there are many hundreds, perhaps even thousands, of different plants in the Southwest mountains and this book describes only about 150 of them. Now you can see that there are several pitfalls in using this book to identify mountain flowers and trees. With practice, however, your misses should become less frequent and your acquaintance with flowers should be greatly increased.

Don't worry about the very few technical terms used — they are all explained in the drawing below showing flower parts.

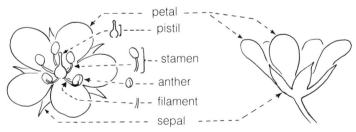

petal
pistil
stamen
anther
filament
sepal

Mountains, plants, and life zones

Most of us know about the high and rugged Rocky Mountains in Colorado, but we are not so familiar with the many beautiful mountains of Arizona, New Mexico, and Utah. Contrary to popular opinion, Arizona and New Mexico contain large areas of forest-covered mountains as well as hot, dry deserts. The same may be said of Utah. The highest elevation in these four states is in Colorado, where Mt. Elbert towers to a height of 14,431 feet. Kings Peak in Utah reaches 13,498 feet, Wheeler Peak in New Mexico, 13,151 feet, and in Arizona the highest elevation is 12,670 foot Humphreys Peak in the San Francisco Mountains. In addition to these typical mountains there are unlimited high plateau areas which reach a sufficiently high elevation to support mountain flora.

There are several kinds of mountains in the Southwest, and it may be surprising to find that not all are snow-covered, craggy peaks. Few would realize they were on top of a mountain when they traveled across the rolling, heavily forested Kaibab Plateau of northern Arizona. This is a huge, domed-up plateau with gently sloping sides reaching an elevation of well over 9,000 feet on top, which is high enough to give it mountain standing in most leagues. Another kind of mountain results from volcanic activity which may build up a pile of rock to a height of several thousand feet. Such a mountain is illustrated by the San Francisco Peaks of northern Arizona. Other kinds of mountains are produced by movements deep in the earth's interior, which cause huge sections of the earth's crust to be shoved up into a mountain range such as the Rockies of Colorado or the Wasatch Range of Utah. In some cases these crustal movements result in formation of rather isolated mountain masses such as the Santa Catalinas of southern Arizona or the Sacramento Mountains of New Mexico.

Not only do we determine the presence of a mountain by the elevation and rugged topography, but also by the kind of plants we find. Some time ago scientists discovered that the surface of the earth could be divided into life zones, each zone containing a characteristic type of plant and animal life. The range of these life zones is determined largely by latitude and elevation, and there are seven of them listed below, with their typical Southwestern plants. (Also see chart on page 2):

Tropical — not represented in southwestern United States.
Lower Sonoran — saguaro, mesquite, catclaw, and paloverde.
Upper Sonoran — pinyon and various species of juniper.
Transition — ponderosa pine.
Canadian — fir and spruce.
Hudsonian — spruce, bristlecone pine, and limber pine.
Arctic-Alpine — above timberline and no trees; other plants small and often matlike.

This book is concerned primarily with plants growing in the last four life zones.

To see all of these zones at sea level it would be necessary for you to go from near the equator to the far northern parts of Alaska or Canada, a distance of many thousands of miles. However, by climbing a mountain it is possible to go through these same zones of life in only a few miles. This is well illustrated by taking the area around Grand Canyon National Park as an example. Starting in the bottom of the canyon we find ourselves in a warm, dry desert supporting vegetation typical of the Lower Sonoran Zone of Mexico and southern Arizona, but as we climb upward we come to a pinyon-juniper woodland typical of the Upper Sonoran Zone. On the canyon's south rim we are in a ponderosa pine forest which denotes the Transition Zone, but only ten miles away as the crow flies, on the north rim, which is about 1,000 feet higher than the south rim, we are in a cool, moist forest typical of the Canadian Life Zone. To find the last two zones we need only go about 50 miles to the south where the Hudsonian Life Zone is represented on the upper slopes of the San Francisco Peaks, and above timberline the Arctic-Alpine Zone is indicated by vegetation that is somewhat similar to that growing in the Arctic regions of the far north.

There are several reasons for this variation in living things as we go either northward or upward. Most important of these is temperature, for the higher we go the cooler it becomes. That is usually the reason for going to the mountains for our summer vacation. A good rule of thumb to remember is that for every increase of 1,000 feet there is a drop in temperature of about 3 degrees Fahrenheit. Interestingly enough, going up 1,000 feet is roughly equivalent to going north about 300 miles at sea level. Of course, the mountains receive more moisture in the form of both rain and snow than the surrounding lowlands and this has a great effect upon the kinds of plants that grow there.

Although the plants found in the different life zones on mountains are similar, they are not always exactly the same as the plants growing in the corresponding zones at sea level. Don't interpret the life zones too rigidly, for many other factors such as soil, slope, exposure, humidity, etc., have a tremendous effect upon the kind of vegetation that grows in any area.

The National Parks and Monuments as wildflower sanctuaries

The National Park Service was created by act of Congress in 1916, and, with respect to the Parks and Monuments, was given the responsibility to "conserve the scenery, scientific and historic objects and the wildlife therein; and to provide for their enjoyment in such a way and by such means as will leave them unimpaired for the enjoyment of future generations." There are today more than 300 units administered by the National Park Service throughout this country including Alaska and Hawaii. With the increasing inroads of various kinds of exploitation into the beauty spots of this country, I'm sure it is comforting for most of us to know that through the National Park Service there will always be preserved a little bit of the finest of primitive and unspoiled America.

Part of the scenery under protection in the Parks and Monuments consists of the native vegetation. Only in such areas are the plants protected from grazing, competition of exotics, and from the destructive activities of mankind. Therefore, the National Parks and Monuments are among the best of places to see and enjoy wildflowers.

Identification Chart for Evergreens

Pines

needles in bundles
with thin "sheath"
holding needles
together

cone scales
thick

cones woody

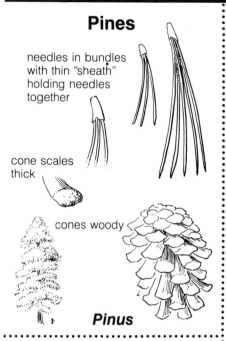

Pinus

Spruces

needles single
sharp
stiff
square
□

twigs
rough
after needles
fall off

cone-scales thin,
papery

cones always
hang down

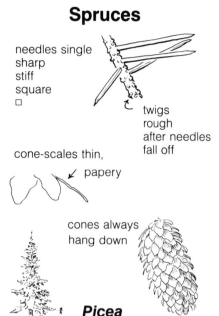

Picea

Firs

needles single
flexible
blunt and flat

twigs with
smooth
round scars
after needles
fall off

cones always erect

central axis of cone
stays after scales
drop off

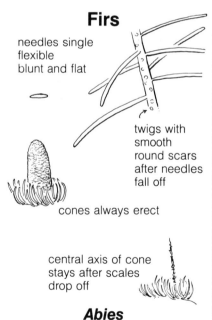

Abies

Douglas-Fir

needles single
flat and
narrowed
at base

identifying feature the
3-pointed bract on cone

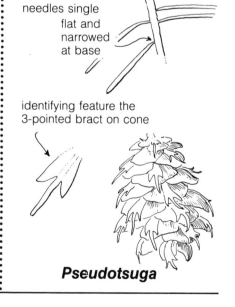

Pseudotsuga

bristles on cone scales

natural size

leaves in 5's

Bristlecone Pine

Pinus aristata

Pine family

9,000' - 12,000'
Colorado, New Mexico to n. Arizona

At very high elevations the bristlecone pine may be almost shrublike, while in more favorable locations it may be a bushy-crowned tree up to 40 feet high. The cones are quite distinctive, as each scale is tipped with a sharp, slender, bristlelike prickle. The needles grow in bundles of five, and are usually about 1¼ inches long.

Pinus longaeva commonly called Bristlecone Pines found in Nevada's White Mountains, 4,000 to 6,000 years old, are considered the oldest living things on earth.

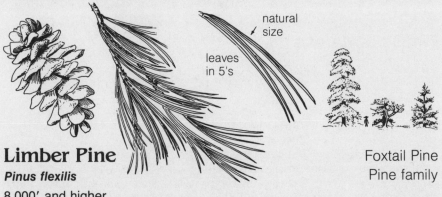

natural size

leaves in 5's

Limber Pine

Pinus flexilis

Foxtail Pine
Pine family

8,000' and higher
Alberta to w. Texas, Arizona to California

Here is another pine to look for as you drive or hike through the high country of the Southwest. It may be distinguished from bristlecone pine, with which it is often associated, by the lack of bristles on the cone scales and by the longer needles.

Large trees may reach a height of 50 to 60 feet, and the long branches often bend gracefully toward the ground. The limber pine has only slight commercial value.

Ponderosa Pine

Western Yellow Pine

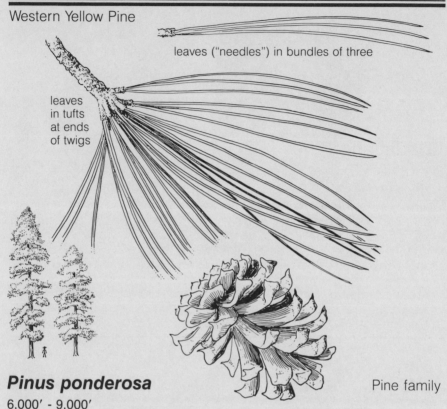

leaves ("needles") in bundles of three

leaves
in tufts
at ends
of twigs

Pinus ponderosa

Pine family

6,000' - 9,000'
British Columbia to Mexico and all the Pacific Coast and Rocky Mountain states

Throughout the Southwest ponderosa pine is the most abundant species of pine. It is the characteristic tree of the Transition Life Zone, where it grows in nearly pure stands in open parklike forests.

It is a handsome tree reaching a height up to 150 feet, and its scaly bark is cinnamon-brown to orange-yellow in color. Trees less than 80 to 100 years old have bark that is dark brown or nearly black, and because of this are often called blackjack pine.

Ponderosa pine requires lots of sunlight throughout its life, especially in older age. Dense stands of seedlings may be seen for the first 10 or 15 years, but they thin out rapidly due to increased need for sunlight.

In Arizona and to a lesser extent in New Mexico the lumber industry depends almost entirely upon the ponderosa pine for wood used in construction, interior finish, crating and boxing. The seeds are an important food for squirrels, chipmunks, and birds.

The young twigs give off a spicy orange peel odor when crushed.

when spruce leaves fall off, rough scar remains

leaves 4-angled ☐

natural size

spruce cones always hang down, and occur at top of tree

Engelmann Spruce

Picea engelmannii

Pine family

8,000' - 12,000'
British Columbia to New Mexico, Arizona and the Pacific Coast states

A prominent tree at the higher elevations where it may grow in either pure stands or mixed with white fir, subalpine fir or limber pine. Large trees may reach a height of 100 feet, but near timberline the tree may take the form of a prostrate shrub.

A forest of Englemann spruce usually contains trees of all ages and sizes, as the young trees are able to grow even in the dense shade of the larger trees. Because of its shallow root system this spruce is often blown over by high winds.

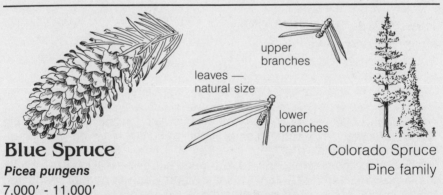

upper branches

leaves — natural size

lower branches

Blue Spruce

Picea pungens

Colorado Spruce
Pine family

7,000' - 11,000'
Rocky Mountain states, Arizona and New Mexico

Blue spruce and Englemann spruce often grow together, and are quite difficult to tell apart. In addition to having larger cones, the crushed needles of blue spruce do not have as disagreeable an odor as those of Engelmann; the twigs are smooth and clean in contrast to the slightly fuzzy twigs of Englemann spruce.

Douglas-Fir

Oregon Pine, Douglas Spruce

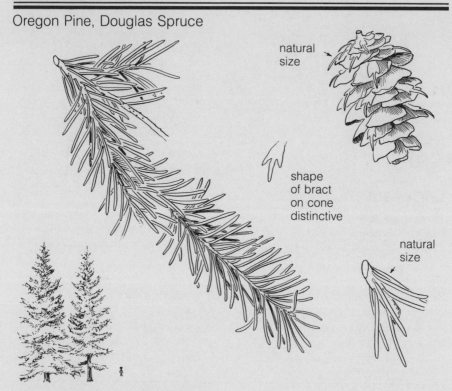

natural
size

shape
of bract
on cone
distinctive

natural
size

Pseudotsuga menziesii Pine family

6,000' - 10,500'
British Columbia, Coast states, Rocky Mountain area, Arizona, New Mexico and n. Mexico

For many years after its discovery on Vancouver Island in 1791, the Douglas-fir was a botanical puzzle. The tree bears a strong resemblance to spruce and fir, as well as to hemlock and yew. For a long time it bore the name *Pseudotsuga taxifolia* which is Greek and Latin, meaning "false hemlock with a yewlike leaf."

Excepting giant sequoias and redwoods of California, Douglas-fir is far the largest western forest tree. In Oregon and Washington it exceeds 300 feet in height, but is much smaller in the Southwest. In favorable locations, however, it is probably the largest tree in the region.

One of the most important timber trees of the Pacific Northwest, it is less important in the Southwest, seldom occurring in pure stands and commonly mixed with ponderosa pine or with spruce at high elevations.

The cones, with the distinctive three-pointed papery bracts, provide the most reliable means of identifying the tree.

Firs

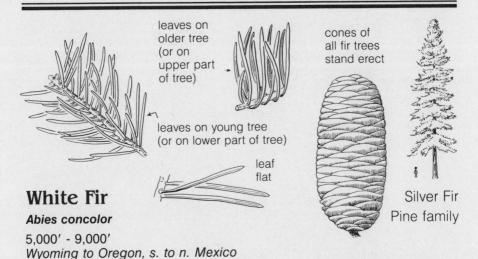

leaves on older tree (or on upper part of tree)

cones of all fir trees stand erect

leaves on young tree (or on lower part of tree)

leaf flat

Silver Fir
Pine family

White Fir

Abies concolor

5,000' - 9,000'
Wyoming to Oregon, s. to n. Mexico

Of the nine different North American firs, the white is probably the most important species. In Pacific Coast States the trees reach 200 feet in height, but Rocky Mountain specimens seldom exceed 100 feet.

It is a fairly rapid grower, reaching maturity in about 300 years. Many of the larger trees are wind-thrown due to the shallow root system. It requires less moisture than other firs, and does well on poor dry sites.

The wood is used largely for construction and cabinet work, and the tree is often planted as an ornamental.

slender tip →

on fir trees, when leaves, (needles) drop off, smooth round scars remain

Alpine Fir
Pine family

Subalpine Fir

Abies lasiocarpa

8,000' and up
Alberta to Alaska and s. to Arizona, New Mexico and Oregon

True to its name, the Alpine fir grows in the high and cool Canadian and Hudsonian Life Zones. The purplish, erect cones are from 2 to 4 inches long, and they often drip with a silvery resin in warm weather.

The tree is relatively small and unimportant, but is the most widely distributed fir in western North America.

Common Juniper

Dwarf Juniper

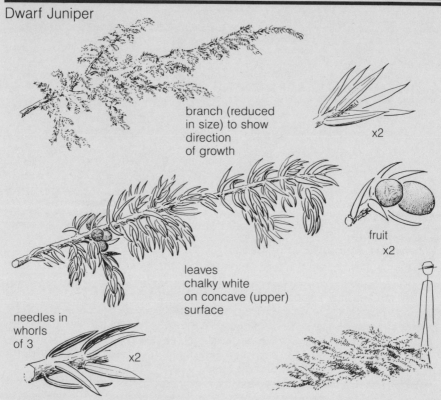

branch (reduced
in size) to show
direction
of growth

x2

fruit
x2

leaves
chalky white
on concave (upper)
surface

needles in
whorls
of 3

x2

Juniperus communis

Cypress family

8,000' - 12,000'
Cooler parts of North America and in Europe and Asia

Although not a tree, this plant is placed with them, for it bears no obvious flowers which would allow it to be placed according to flower color. Its relatives, alligator juniper, Rocky Mountain juniper, and other species all reach tree size.

Common juniper is usually less than 3 feet high, and forms a straggling shrub. The leaves consist of rather loose-fitting, sharp-pointed scales, and the plant bears large bluish berries.

Indians of various parts of the country have made use of the berries and other plant parts for food and beverages. The berries are relished by birds and animals, and are used commercially to give flavor to gin and to prepare oil of juniper, used in patent medicines.

Whenever it is abundant, common juniper forms a valuable ground cover for controlling erosion of the soil. This is of particular value at the higher elevations where vegetation and ground cover is scant.

Quaking Aspen

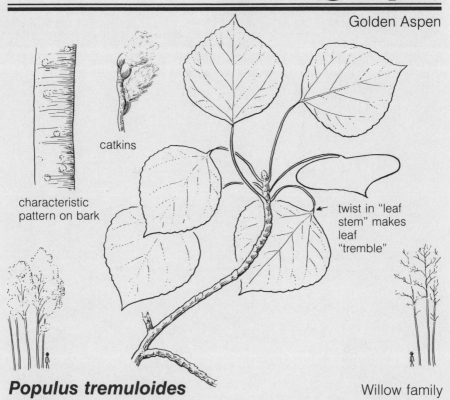

catkins

characteristic pattern on bark

twist in "leaf stem" makes leaf "tremble"

Populus tremuloides

Willow family

6,500' - 9,500'
High areas of the west, throughout Canada, n. and ne. United States

One of the most beautiful sights to be seen in the western mountains during the fall is the quaking aspen with its brilliant golden leaves and white powdery bark. Aspens form a veritable belt of gold around the San Francisco Peaks in northern Arizona which may be seen from miles away. It is a memorable experience to walk through an aspen grove at this time of year while the golden leaves are falling like snowflakes, building up a yellow carpet upon the ground, which will soon be covered by winter snows.

During the summer the leaves are shiny green on the upper surface and a pale dull green below. The tree is well named, for the slightest breeze will cause the delicately balanced leaves to tremble and dance.

Aspens often grow on burned or cut-over areas, where they help in preventing erosion. The wood is used for paper pulp, boxes, excelsior, and matches.

Trees reach a height of 80 feet and a diameter of 30 inches, but are usually much smaller. Because of the similarity in bark, aspens are often confused with birch.

15

Rocky Mountain Maple

Dwarf Maple

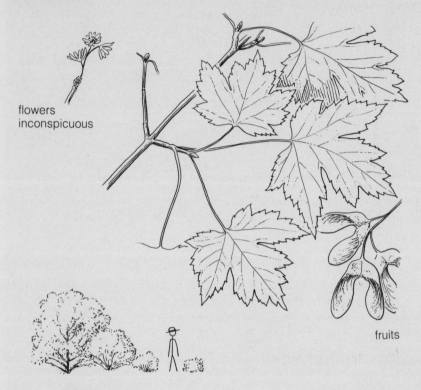

flowers
inconspicuous

fruits

Acer glabrum   Maple family

5,000' - 8,500'
South Dakota to Alaska, s. to New Mexico, Arizona and California

The beauty of Rocky Mountain maple is most easily appreciated in the autumn when it splashes mountainsides with rich reds and bright yellows. Maples, including the common boxelder, are quite attractive, but are of minor importance in the Southwest.

Growing mostly in damp woods in the Transition and Canadian Life Zones, Rocky Mountain maple may reach a height of 30 feet, but is usually smaller and shrubbier. The leaves are three- to five-lobed, have toothed edges, and the leaf stems and buds are reddish. Keep in mind the trademarks of the maple family which include opposite leaves and fruit consisting of a pair of winged seeds united at the base.

Our western maples are of little economic value when compared to the maples of the east. Deer and livestock browse the plants, and maple sugar is sometimes made from the sap of the bigtooth maple.

Ponderosa pine, *Pinus ponderosa*

Aspen, *Populus tremuloides,* and white fir, *Abies concolor*

Stephen Trimble

Aspen, *Populus tremuloides,* and penstemon

Stephen Trimble

Kinnikinnick, *Arctostaphylos uva-ursi*

John Richardson

Cranesbill, *Geranium richardsonii*

Jeanne Janish

Elkslip, *Caltha leptosepala*

Stephen Trimble

Chokecherry *Prunus Virginiana*

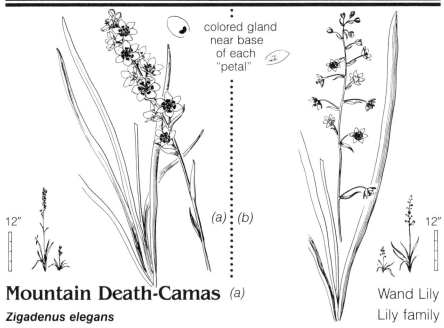

colored gland
near base
of each
"petal"

12"

(a) (b)

12"

Mountain Death-Camas *(a)*

Zigadenus elegans

Blooms yellowish-white. July-August. 5,000' - 10,000'
Saskatchewan to Alaska, s. to New Mexico and Arizona

Wand Lily
Lily family

Mountain death-camas, with cream-colored flowers grouped at the top of straight, slender stems, may often be found growing in moist, cool places in the mountains. The stems, from 6 inches to 3 feet tall, spring from a clump of stiff bluish green leaves. Flowers are usually about one-half inch in diameter, and at the base of each petal is located a small spot or gland.

This plant is somewhat poisonous, but less so than other members of the genus, which sometimes cause heavy loss of sheep and cattle. The toxic agent, called zygadenin, appears to be in all parts of the plant, even in the seeds.

Death-Camas *(b)*

Zigadenus virescens

Lily family

Blooms greenish-white. July-September. 6,500' - 11,000'
New Mexico and Arizona to central America

Although not as widely distributed as mountain death-camas, this death-camas may be found growing in the rich soil of coniferous forests of southern Arizona and New Mexico. It closely resembles *Zigadenus elegans,* but the flower head is more spreading and the individual flower stems, called pedicels, have a tendency to curve away from the main stalk.

Death-camas grows from bulbs which somewhat resemble onions.

California False-Hellebore

Skunkcabbage

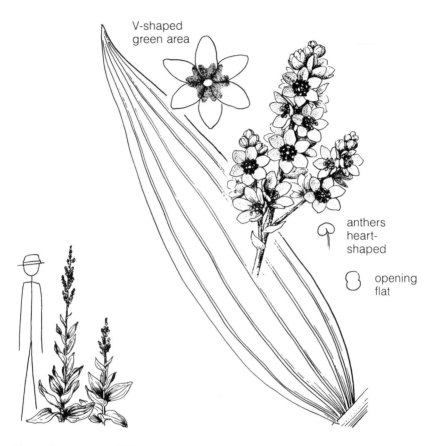

V-shaped green area

anthers heart-shaped

opening flat

Veratrum californicum

Lily family

Blooms creamy-white, streaked with green. July-August. 7,500' - 9,500'
Montana and Washington, s. to New Mexico, Arizona and California

A tall coarse plant from 3-6 feet tall. Its most conspicuous features are the bright yellowish-green leaves which are broad and rounded with heavy ribs, giving them a pleated appearance. Flowers are borne along the upper part of the stem, and are about one-half inch across.

A poisonous substance, veratrin, is contained in the root and young shoots, and may poison stock, although the plant is seldom eaten. Flowers are poisonous to insects and heavy losses in honeybees sometimes occur.

The powdered roots of some species of false-hellebore are used in making insect powder.

Solomon-Plumes

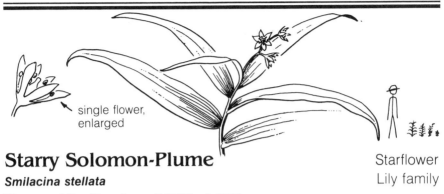

single flower, enlarged

Starry Solomon-Plume
Smilacina stellata

Starflower
Lily family

Blooms white. May-June. 7,000' - 9,000'
Found throughout most of temperate North America and in Europe

This plant and the next one grow in very similar situations and, in fact, are sometimes found side by side. Starry Solomon-plume bears green berries with dark vertical stripes which turn black upon maturity.

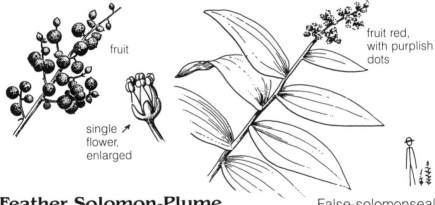

fruit

single flower, enlarged

fruit red, with purplish dots

Feather Solomon-Plume
Smilacina racemosa

False-solomonseal
Lily family

Blooms white. May-July. 6,000' - 10,000'
Grows throughout most of temperate North America

You may expect to find this plant almost anywhere in the Transition and Canadian Life Zones where there are cool shaded places and rich soil. It may reach 2-3 feet in height, and the tall flowering stems branch from horizontal rootstalks.

The flowers are very small and are creamy white. After blooming, small red berries with purplish dots develop. The plant may be distinguished from starry solomon-plume by its broader leaves, smaller berries, and more abundant flowers.

Elkslip

Marshmarigold

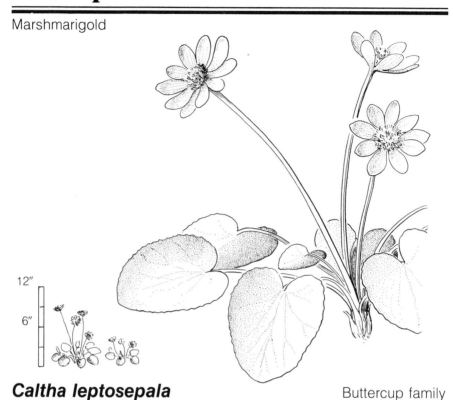

12"

6"

Caltha leptosepala   Buttercup family

Blooms white. June-September. 9,000' - 11,000'
Montana to Alaska, s. to New Mexico, Arizona and Washington

This is truly a plant of the high mountains — it may often be found blooming at the edge of melting snowbanks or other such marshy areas. The flowers are quite handsome, and up to 1¼ inches across.

Odd as it may seem, the flowers have no true petals at all. The large white petal-like parts of the flower, which are tinged with bluish on the under side, are actually sepals. The flower is yellow-centered due to the presence of a great many small, yellow structures called anthers.

The plants may reach a height of about 8 inches, and have stout purplish stems. The leaves, growing mostly at the base of the plant, are light green and often have purple veins on the under side.

Quite often these pretty little plants are found growing on tiny, boggy islands in mountain brooks and streams, sometimes growing right in the water and rooted to the bottom.

Another species of marshmarigold, *Caltha palustris,* was used for food by the Indians in Minnesota, Wisconsin, and the eastern States. Its leaves and stems were boiled for greens. However, when raw they are irritating.

Pennycress

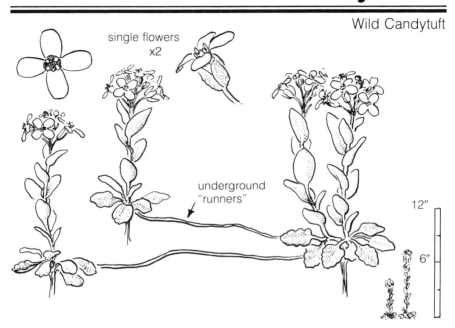

single flowers
x2

underground
"runners"

12"

6"

Thlaspi montanum var. *fendleri*   Mustard family

Blooms white. February-August. 4,000' - 12,000'
Colorado, Utah, New Mexico and Arizona

The small white flowers of "wild candytuft" are far from being showy, and the plant itself is not conspicuous, as it is only a few inches tall. However, it is a very common plant in coniferous forests and due to the fact that it blooms so early, when few if any other plants are in flower, it is quite noticeable.

Even when the snow is still on the ground, the little white flowers of "wild candytuft" may sometimes be seen. After a long hard winter they are a very welcome sight, for their appearance is a good sign that spring is just around the corner.

There are many kinds of *Thlaspi* growing in temperate and artic regions, mostly in the mountains. As a group they are generally smooth, low plants with the basal leaves arranged in a rosette on the ground, while the stem leaves are small and tend to clasp the main stem.

This plant, like all other members of the mustard family, has flowers with four petals. In fact, it is because of the cross-like arrangement of the four petals that the family has been given the scientific name of Cruciferae.

The generic name, *Thlaspi,* is from the Greek word *thlan,* meaning to crush, on account of the strongly-flattened, triangular pods which are borne by pennycress.

Mountain-Ash

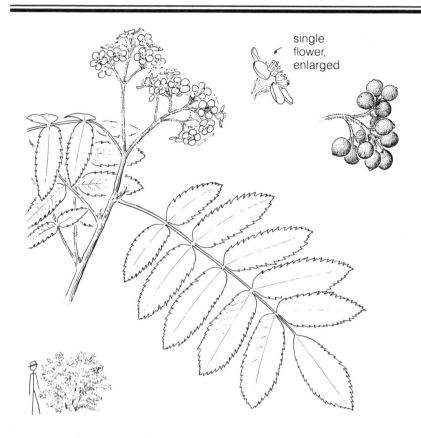

single flower, enlarged

Sorbus dumosa

Rose family

Blooms white. June-July. 8,000' - 10,000'
New Mexico and Arizona

This is a handsome shrub or small tree bearing large flat-topped or rounded clusters of white flowers. Later in the fall the flowers are replaced by brilliant orange-red berries. Under favorable conditions the plant may reach a height of 9 feet.

It should be looked for in the moist rich soil of coniferous forests in the Transition and Canadian Life Zones.

There are several species of mountain-ash which are widely distributed in the mountainous regions of North America. They are, of course, not related to the true ashes which are members of the olive family.

The bark of one of the eastern species is used medicinally for its tonic, astringent, and antiseptic properties. Wild animals and especially birds relish the bright red mature berries.

Western Thimbleberry

Salmonberry

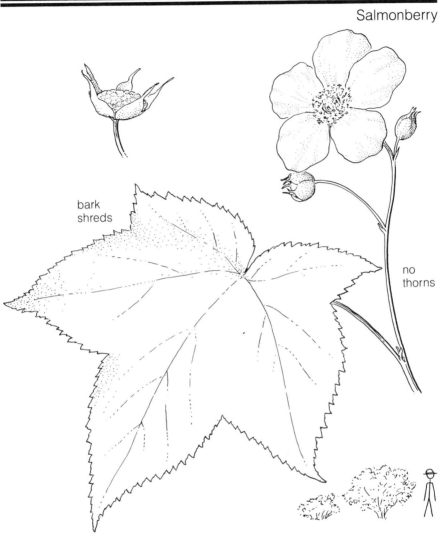

bark
shreds

no
thorns

Rubus parviflorus

Rose family

Blooms white. July-September. 8,000' - 9,500'
Alaska to Michigan, s. to New Mexico, e. Arizona and California

In canyons and on wooded slopes of the ponderosa pine and spruce belts we find this attractive plant. The white flowers measure up to 2 inches across, and grow in clusters of three or more at the ends of long stems.

The fruit is raspberry-like in appearance, but is disappointing to the taste, for it is composed mostly of seeds.

Raspberries

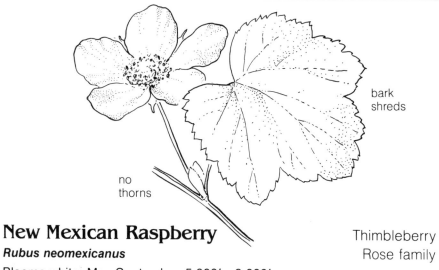

bark
shreds

no
thorns

New Mexican Raspberry
Rubus neomexicanus

Thimbleberry
Rose family

Blooms white. May-September. 5,000' - 9,000'
New Mexico, Arizona and n. Mexico

Not as widespread as western thimbleberry; more abundant in Arizona and southwestern New Mexico. Both species are reported to be extensively browsed by deer.

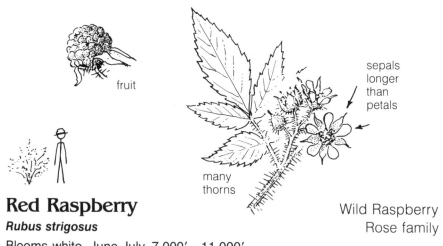

fruit

sepals
longer
than
petals

many
thorns

Red Raspberry
Rubus strigosus

Wild Raspberry
Rose family

Blooms white. June-July. 7,000' - 11,000'
Grows in the cooler parts of North America

This is the granddaddy of some of our cultivated raspberries. Berries are delicious raw or in jam or jelly; are relished by birds and animals.

Common Chokecherry

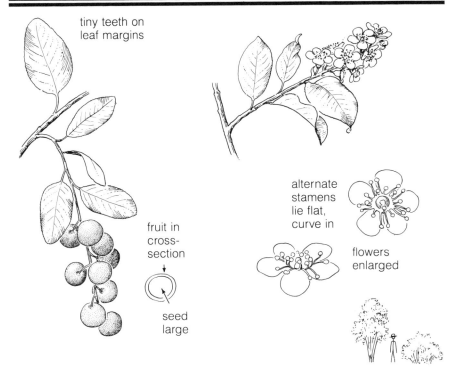

tiny teeth on leaf margins

fruit in cross-section

seed large

alternate stamens lie flat, curve in

flowers enlarged

Prunus virginiana
Rose family

Blooms white. April-June. 4,500' - 8,500'
Canada to Georgia, New Mexico, Arizona and California

Usually a tall shrub, but under favorable conditions the plant may reach a height of 25 feet. It is quite common at the middle elevations where it is often associated with the ponderosa pine.

In addition to being a delicacy to animals, the fruits are gathered by man to be used in making jams and jellies. A purplish-red dye can be extracted from the berries, and in the spring the inner layer of bark yields a green dye.

The chokecherry is a sacred plant to the Navajo. Various dance implements are made of the wood, and it is often mentioned in songs. Prayer sticks of the North are made from the wood, possibly because the fruits ripen black, the tribe's color of the North.

From a horticultural standpoint the genus *Prunus* is one of the most interesting and important of all plant genera. It includes such orchard fruits as peaches, plums, cherries, apricots, and almonds; and many of its members provide such ornamental subjects as double-flowered, variegated-leaved, colored-leaved and weeping forms.

Wild Strawberry

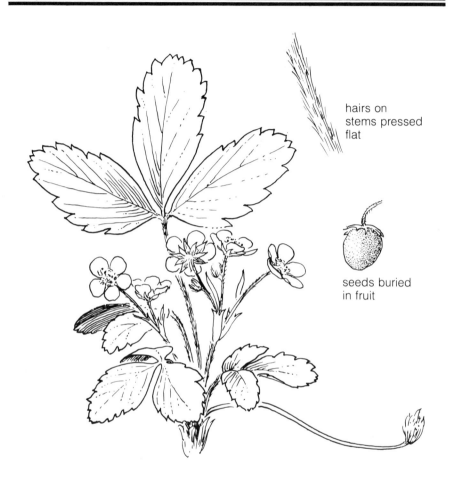

hairs on
stems pressed
flat

seeds buried
in fruit

Fragaria ovalis

Rose family

Blooms white. May-October. 7,000' - 11,000'
Wyoming to New Mexico and Arizona

Most people need no introduction to the wild strawberry, for it looks much the same as the cultivated strawberry grown in gardens. Its leaves, divided into three distinct leaflets, are typical of all strawberries, and it bears an attractive five-petaled flower.

Strawberries spread by runners which root and start new plants. The berries are rather small, but tasty, and are relished by birds and other animals.

Bracted Strawberry

Wood Strawberry

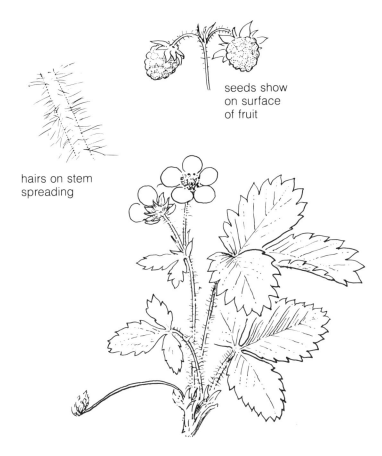

seeds show
on surface
of fruit

hairs on stem
spreading

Fragaria bracteata

Rose family

Blooms white. May-September. 7,000' - 9,500'
Montana to British Columbia, s. to New Mexico, Arizona and California

The white flowers of the bracted strawberry are almost an inch across, and have fuzzy, bright yellow centers. In this species the seeds are only slightly attached to the surface of the berry, while in *Fragaria ovalis,* the seeds are partly buried in the flesh of the berry.

The generic name, *Fragaria,* comes from the Latin, meaning fragrance, in reference to the smell of the fruit.

31

Arizona Peavine

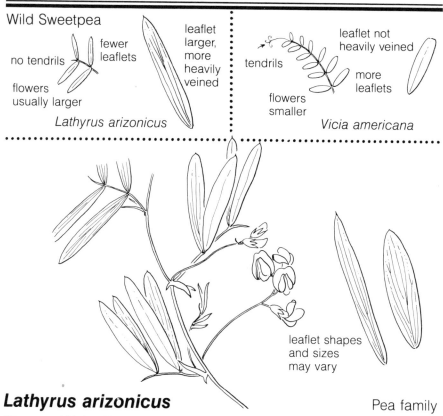

Wild Sweetpea

no tendrils → fewer leaflets

flowers usually larger

leaflet larger, more heavily veined

Lathyrus arizonicus

tendrils

leaflet not heavily veined

more leaflets

flowers smaller

Vicia americana

leaflet shapes and sizes may vary

Lathyrus arizonicus

Pea family

Blooms whitish with pinkish veins. May-October. 8,000' - 11,000'
Colorado and se. Utah to n. central Mexico

Wild sweetpeas greatly resemble the vetches belonging to the genus *Vicia,* but have fewer, larger, and more prominently veined leaflets, and larger flowers that are almost a carbon copy of those borne by the cultivated sweetpea. In addition to these characters, this plant may be distinguished from the vetches at a glance, for its tendrils at the ends of the leaves are greatly reduced and bristlelike. Those of the vetches are long and twining.

The cultivated garden sweetpea, *Lathyrus odoratus,* a native of Sicily, is the best known member of the genus. Altogether the genus contains about 200 species, occurring mostly in the northern hemisphere. Several species are cultivated primarily for their showy flowers, although they are also of value as screening plants, for they quickly cover trellises and other such structures.

The seeds and pods of one of the other southwestern species are used for food by some of the Indians of New Mexico. Although palatable to livestock, the plants are secondary in importance to the vetches.

Richardson Geranium

Cranesbill

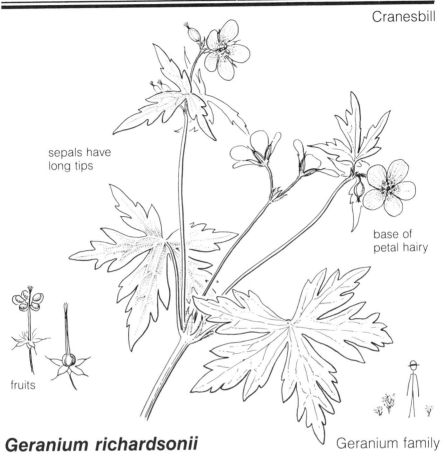

sepals have long tips

base of petal hairy

fruits

Geranium richardsonii

Geranium family

Blooms white with occasional purple tinge. April-October. 6,500′ - 11,500′

Be on the watch for this wild white geranium in the moist rich soil of coniferous forests. It is a tall, slender plant having thin, many-lobed leaves and white flowers with delicate pink or purplish veins.

The geranium family, though small, has contributed many handsome plants to cultivation. Cultivated geraniums bear a resemblance to wild geraniums, but are members of the genus *Pelargonium* and hail from South Africa. There are more than 250 species of geraniums growing in the temperate zones of the northern hemisphere. A very few species grow in the tropics.

Sheep are reported to browse on the plants, and various species are used medicinally as an astringent.

The name geranium is from the Greek word *geranos,* meaning a crane, in reference to the long beak which forms on the seed capsule.

Common Poison-Ivy

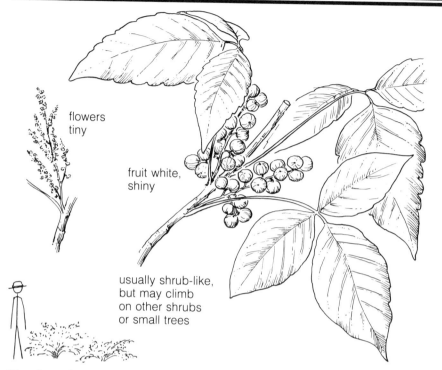

flowers
tiny

fruit white,
shiny

usually shrub-like,
but may climb
on other shrubs
or small trees

Toxicodendron radicans

Cashew family

Blooms whitish to greenish white. April-September. 3,000' - 8,000'
Throughout most of North America

This plant is usually no beauty, although its foliage is handsome in autumn, but for obvious reasons we should make its acquaintance. It is not especially abundant in the high mountains, but is quite common in the rich soil of ravines and foothill canyons.

The toxic agent of poison-ivy is a nearly nonvolatile oil called urushiol. It is contained in all parts of the plant, and causes skin eruptions and swelling in susceptible persons. Even the milky juice is poisonous when taken internally. If you come in contact with the plant, you immediately should wash the affected part with strong soap and water.

All parts of the plant are utilized by various birds and animals for food. In fact, the plant may provide as much as 10 to 25% of the diet of some wild animals. It has always been puzzling to the writer why these animals don't break out with a bad case of pimples after eating such a toxic dinner, but apparently it never happens. The woodpeckers, in particular, must have cast-iron stomachs, for they not only like ants, with their bitter taste of formic acid, but they also relish the berries of poison-ivy.

Fendler Ceanothus

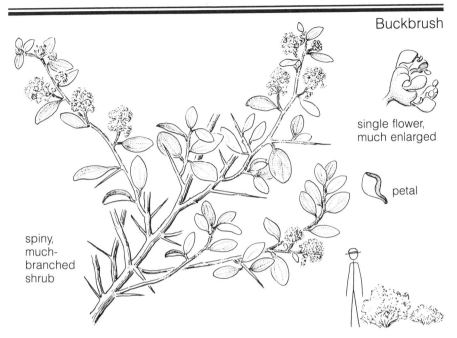

single flower,
much enlarged

petal

spiny,
much-
branched
shrub

Ceanothus fendleri

Buckthorn family

Blooms white. April-October. 5,000' - 10,000'
Colorado and Utah to w. Texas, New Mexico, Arizona and n. Mexico

A low, thorny shrub, very common in pine forests; in fact, it grows almost everywhere except in the desert. Under favorable conditions, the shrubs reach about 3 feet and form thickets. At high elevations, buckbrush may be no more than a foot in height.

The leaves, which are browsed extensively by livestock and deer, are rather thick, about twice as long as wide. They have smooth edges and are usually fuzzy white beneath. The small, white flowers are clustered at the ends of the twigs.

It is said that the flowers will form a lather in water. A good honey is made from the nectar in the small blossoms. In California, buckbrush may make up as much as 50% of the diet of the mule deer.

Other members of this genus have been cultivated as ornamentals, and are sometimes known as wild-lilac. The Navajos make a medicine from ceanothus which they apply externally, and also take internally to relieve nervousness and alarm. Indians of New Mexico are reported to use the berries for food.

Of the 55 species of *Ceanothus* in the world, 4 occur in the East, and 40 grow in the West. The greatest number are found in California and neighboring states.

Canada Violet

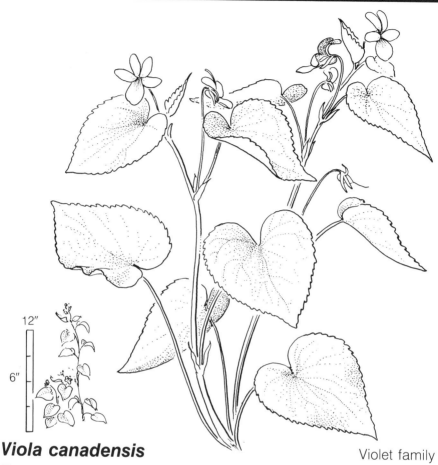

12"

6"

Viola canadensis

Violet family

Blooms white. April-August. 6,000' - 11,000'
Throughout Canada and s. to Alabama, New Mexico and Arizona

Familiar to almost everyone are the various kinds of violets. This one has rather tall, weak stems, often a foot or more in height, bearing smooth, heart-shaped leaves. The flowers are white with purplish veins, and usually tinged with purple on the back. Occasional flowers are entirely white, and sometimes sweet scented.

Although many of the North American species of violets have beautiful flowers, they have not been extensively cultivated. On the other hand, the sweet violet and the pansy, both of which are native to Europe, have long been favorites in flower gardens.

Rich, moist soil in coniferous forests of the Transition and Canadian Life Zones offer the best conditions for the Canada violet.

Porter Ligusticum

Chuchupate, Osha

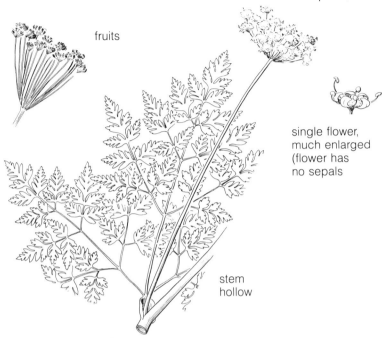

fruits

single flower,
much enlarged
(flower has
no sepals

stem
hollow

Ligusticum porteri

Parsley family

Blooms white. June-August. 6,500' - 11,500'
From Wyoming to Arizona and Chihuahua

This is a rather stout plant reaching a height of 3 feet, and having fern-like leaves. It grows in moist fertile ground almost to timberline and is a palatable forage plant.

In northern New Mexico, the plant is called *osha* by the Spanish-speaking New Mexicans and Indians, who prize it highly for its medicinal qualities. Osha roots, prepared in various ways, are used to treat such disorders as upset stomach, colds, flu, pneumonia, tuberculosis, fevers, cuts, rheumatism, headaches, snakebites and even colic.

A pungent odor is given off by its foliage, which is somewhat similar to that of its relative, celery. In fact, osha leaves are often used like celery in seasoning soups and other dishes. The dried hollow stems were smoked in place of cigarettes by the Apache and, later, by the Spanish settlers. On the other hand, the roots are now chewed in an effort to break the tobacco habit.

The parsley family contains many useful plants such as carrots, parsnips, celery, dill and anise. However, the water hemlock, *Cicuta* spp., and the poisonhemlock, *Conium maculatum,* are among the most poisonous of plants.

37

Common Cowparsnip

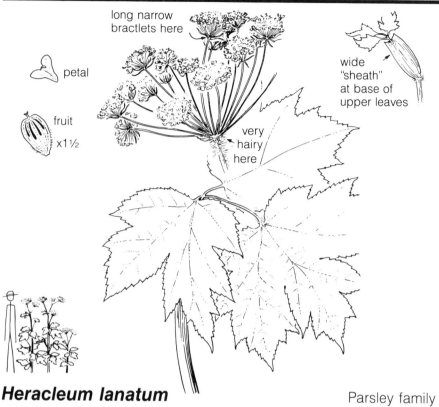

long narrow
bractlets here

petal

fruit
x1½

very
hairy
here

wide
"sheath"
at base of
upper leaves

Heracleum lanatum

Parsley family

Blooms white. July-August. 7,500' - 9,000'
Widely distributed in the United States and Canada

This is one of the showiest members of the parsley family. It is a coarse plant standing from 3 to 6 feet tall with leaves having three large leaflets, irregularly toothed and notched on the edges.

The flat clusters of white flowers are really enormous, being up to 8 inches across. The scientific name, *Heracleum,* refers to Hercules because of the great size of the plant. It often grows along streams, or in meadows in the ponderosa pine and spruce belts.

Many Spanish-speaking New Mexicans call the plant *yerba del oso,* meaning bear weed. They use the powdered root in many ways, such as rubbing it on the gums when teeth are loose, applying it to the body to reduce fever, and bathing parts of the body afflicted with rheumatism.

Besides its medicinal properties, the plant is relished by livestock; the tender leaves and stems were eaten by Indians. It is stated that the cooked roots taste like rutabaga. Susceptible persons may suffer skin eruptions as a result of touching the wet foliage.

Redosier Dogwood

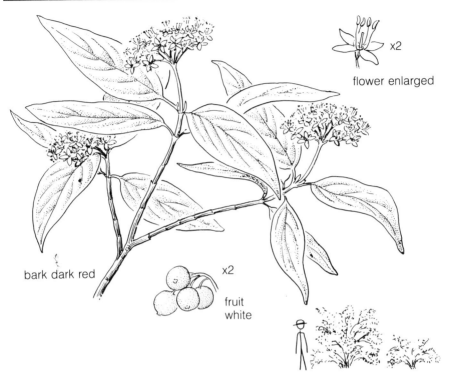

flower enlarged

x2

bark dark red

x2

fruit
white

Cornus stolonifera   Dogwood family

Blooms white. May-July. 5,000' - 9,000'
Canada and Alaska, s. to the District of Columbia, New Mexico, Arizona and California

Shrubs from 3 to 8 feet high, common along streams and often associated with willows. The branches have smooth, dark red bark, while the bark on the twigs is a brighter red. Leaves are rather thin, smooth, dark green above, and somewhat whitish below.

Handsome, flat-topped clusters of small, creamy-white flowers are borne on the plant. These clusters may be 2 or more inches in diameter. After the flowers have fallen, small, white, berrylike fruits develop, which are considered a delicacy by birds and animals.

The genus receives its name from the Greek word for horn, because of the toughness of the wood. The bark of *Cornus sanguinea* was once used in England to wash mangy dogs and the name dogwood stems from this.

The bark of the flowering dogwood contains a useful substitute for quinine, and it is said that it is sometimes possible to ward off fevers by merely chewing the twigs.

Woodland Pinedrops

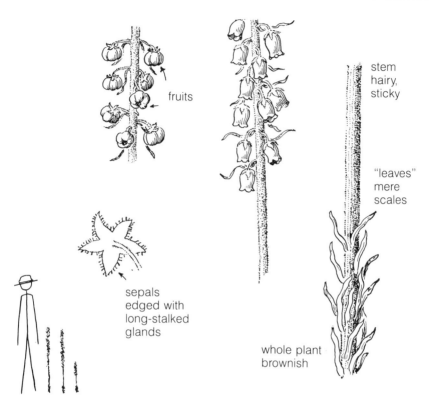

fruits

stem hairy, sticky

"leaves" mere scales

sepals edged with long-stalked glands

whole plant brownish

Pterospora andromedea   Heather family

Blooms ivory-white to yellowish. June-August. 6,000' - 9,500'
Canada to Pennsylvania, Arizona, California and n. Mexico

Usually growing in coniferous forests across the continent, but nowhere really common, the pinedrops is a strange and interesting plant. As many of us know, most plants manufacture their own food through a process called photosynthesis, but this is true of only those plants having green leaves or stems.

Take a close look at the pinedrops, and you will notice that it has no green parts at all. Because of this lack of green stuff, called chlorophyll, it cannot provide its own nourishment, and so it must live on dead organic material. This kind of plant is called a saprophyte, while those plants such as mistletoe, living on live organic material, are called parasites.

The plant itself consists simply of an unbranched, fleshy stem from 1 to 3½ feet high, along which hang many small bell-like flowers. The stem is usually brownish or reddish in color, and is rather sticky.

Pyrolas

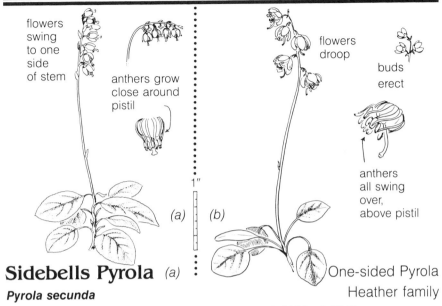

flowers swing to one side of stem

anthers grow close around pistil

flowers droop

buds erect

anthers all swing over, above pistil

1"

(a)   (b)

Sidebells Pyrola *(a)*
Pyrola secunda

One-sided Pyrola
Heather family

Blooms white to greenish-white. July-August. 7,000' - 9,500'
Widely distributed in the northern hemisphere

Some botanists place this little plant in the genus *Pyrola,* but one of the more recent botanical publications places it in a separate group.

It is a small plant from 3 to 6 inches high with dainty, bell-like flowers arranged on one side of the stem. The leaves are mostly at the base of the plant and have small teeth along their edges.

These plants are sometimes called "shinleaf," as the leaves are used for plasters by English peasants. Another common name is wintergreen, but they should not be confused with the aromatic wintergreen belonging to the genus *Gaultheria.*

Green Pyrola *(b)*

Pyrola virens   Heather family

Blooms greenish-white. July-August. 6,500' - 10,000'
Canada to the District of Columbia, New Mexico, Arizona, California and also in Europe

Similar in appearance to the previous species, but the small, drooping flowers are not arranged on one side only. This, and other species of *Pyrola* should be looked for in moist coniferous forests.

The Kayenta Navajos are reported to use the plant for the treatment of diarrhea in infants and for stopping hemorrhage. It is also combined with red ochre and hematite in a paint used in ceremonials.

Woodnymph

Single Delight

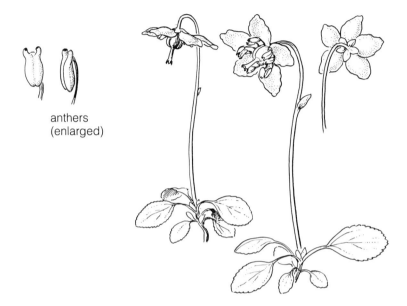

anthers
(enlarged)

Moneses uniflora   Heather family

Blooms white. July-August. 9,500' - 11,500'
Widely distributed in the cooler parts of the northern hemisphere

   This is one of the most attractive of the Southwestern mountain plants, and the search usually necessary to find it will be well worth the time. It is a plant of the moist deep woods, and shows a decided preference for dense shade. Look especially on mossy banks, for such places are sometimes almost covered with its waxy-white flowers.

   The woodnymph is sometimes considered a member of the genus *Pyrola,* and a glance at the plant will show a decided resemblance to the pyrolas. However, the woodnymph has but a single flower usually with five widespread white petals, while those of pyrolas are somewhat closed, making the flowers bell-shaped.

   The single flower stalk is from 1 to 6 inches tall, and springs from a clump of glossy, bright green leaves with toothed edges. The dainty, solitary, sweet-scented flower is about three-fourths of an inch across, and is turned face down, making it almost necessay to pick it to fully appreciate its beauty.

   The generic name *Moneses* comes from the combination of two Greek words meaning "single delight."

   Indians of Montana and Alaska reportedly use the fruit for food.

Swertia

Green-gentian, Deers-ears

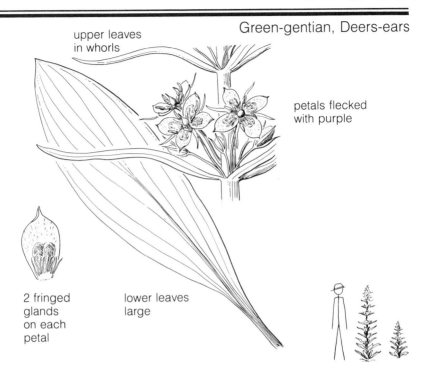

upper leaves
in whorls

petals flecked
with purple

2 fringed
glands
on each
petal

lower leaves
large

Swertia radiata

Gentian family

Blooms greenish-white flecked with purple. May-August. 5,000' - 10,000'
South Dakota to Washington, s. to New Mexico, Arizona, California, ne. Mexico

Among the most conspicuous plants of the mountains is the swertia. It may reach a height of 6 feet, and bears numerous, rather dull colored star-shaped flowers. The pale green leaves are arranged in whorls of four and six, and are usually large and broad at the base of the plant, and become increasingly small and narrow toward the top.

Many insects in search of honey visit the flowers, which bloom all summer. Favored habitats for the swertias include rich soil in open pine forests, aspen groves, and on up into the spruce belt.

In some of the small, isolated native villages in New Mexico, the plant is highly prized for its medicinal properties. A mixture of warm water and powdered root is applied externally to reduce fever. Large doses of the powdered root have proved fatal, but when taken in small quantities it is an effective laxative. A mixture of lard and coarsely ground roots is sometimes applied to the head for killing lice.

The roots of this plant are said to be used for food by the Apache Indians. (This plant has long been known by the botanical name of *Frasera speciosa*).

Pedicularis

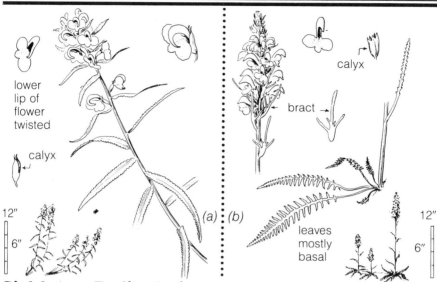

lower lip of flower twisted

calyx

calyx

bract →

leaves mostly basal

12″ 6″

(a) (b)

12″ 6″

Sickletop Pedicularis (a)

Pedicularis racemosa

Woodbetony, Ram's Horn

Snapdragon family

Blooms white to cream. July-August. 9,500′ - 11,000′
Canada to New Mexico, e. Arizona and California

In deep forests of the Hudsonian Zone, sickletop pedicularis is usually fairly common. Stems may be 15 or more inches high, bearing long, narrow leaves with finely toothed margins. Attractive two-lipped flowers grow at tops of stems, and the upper lip is long and curved downward.

The generic name of this plant refers to the Latin word *pediculus,* meaning louse, and doubtless originated because the seeds were used in ancient times to kill lice. This would also explain the common name "lousewort," which is often applied to the plants in this genus. The louseworts are thought to be partially parasitic on the roots of other plants.

Parry Pedicularis (b)

Pedicularis parryi

Parry Lousewort

Snapdragon family

Blooms cream. June-September. 7,500′ - 12,000′
Wyoming to Montana, s. to n. New Mexico and n. Arizona

May be distinguished at a glance from the preceding species by its fernlike leaves and slightly smaller flowers, having the upper lip more open and hood-shaped. The stems may reach a height of 20 inches.

The genus contains about 250 species in various parts of the northern hemisphere, many of them arctic and alpine. Thirty or 40 are probably native in the United States.

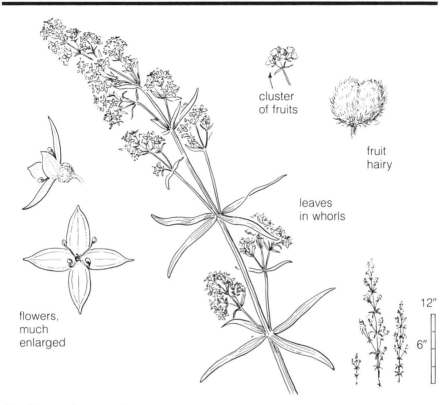

cluster
of fruits

fruit
hairy

leaves
in whorls

flowers,
much
enlarged

12"

6"

Galium boreale
Madder family

Blooms white. July-September. 6,000' - 9,500'
Canada to Pennsylvania, Texas, e. Arizona and California

An attractive plant about 15 inches high bearing dense clusters of small white flowers. The three-veined leaves are arranged in fours in a whorl around the square stem.

The generic name comes from the Greek, meaning milk, as certain species have been used to curdle milk. Some species have been used for medicinal purposes, but with questionable results.

The madder family is quite large, and mostly tropical in distribution. It contains such important plants as the coffeetree, and the trees from whose bark quinine is obtained. Some of the tropical shrubs belonging to this family have nodules on their leaves in which nitrogen-fixing bacteria live. These bacteria are closely related to those found in root nodules of members of the pea family.

There is a tradition that the manger of the Christ Child was filled with these plants; thus the name bedstraw.

Elders

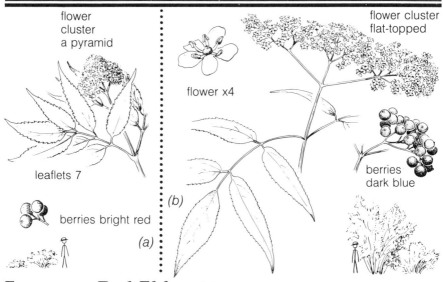

flower cluster a pyramid

flower x4

flower cluster flat-topped

leaflets 7

berries dark blue

berries bright red

(b)

(a)

European Red Elder (a)
Sambucus racemosa

Red-berried Elder
Honeysuckle family

Blooms white to yellowish. June-July. 7,500' - 10,000'
Cooler parts of North America, Europe and Asia

In June these low shrubs are a mass of rounded clusters of small, white flowers which are followed by numerous shiny, scarlet berries in late August. The compound leaves are opposite, and commonly consist of seven leaflets with sharply toothed edges.

Except for this species the fruits of elders are edible and are attractive to birds and some rodents. Wine and jelly are often made from the berries of other species. Cases of poisoning have been reported from eating the berries, flowers, roots, and bark raw.

Blueberry Elder (b)
Sambucus glauca

Honeysuckle family

Blooms cream. June-September. 6,500' - 8,500'
W. Canada to Arizona and s. California

This is the common elder that is so abundant on both rims of Grand Canyon. It is a large shrub up to 20 feet high, having many stems which sprout from the root, and often grows in clumps.

With its large, flat-topped clusters of whitish flowers, it is a truly attractive plant. The branches are pithy in the center, and may be easily hollowed. Elder plants have long been regarded as having medicinal virtues, but these are rather doubtful, although the flowers are reported to be diuretic.

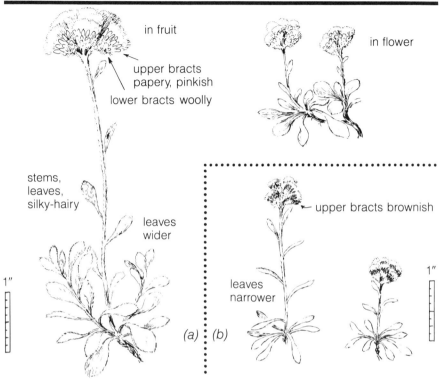

in fruit

upper bracts
papery, pinkish

lower bracts woolly

in flower

stems,
leaves,
silky-hairy

leaves
wider

upper bracts brownish

1″

leaves
narrower

1″

(a) (b)

Rocky Mountain Pussytoes (a)

Antennaria aprica   Sunflower family

Blooms whitish to pinkish. May-August. 5,000′ - 12,000′
Manitoba to British Columbia, s. to New Mexico and Arizona

Looking very much like little balls of fur, the flowers of this plant are rather inconspicuous, but quite common in the Southwest mountains. The flowers are borne on short stems, from 2 to 4 inches high, which arise from mats of whitish foliage.

It is reported that a tobacco made from this plant is smoked by the Navajos in prayer for rain. Parts of the plant are chewed with deer or sheep tallow as a blood purifier.

Pussytoes (b)

Alpine Catspaw

Antennaria umbrinella   Sunflower family

Blooms white to brownish. July-August. 10,000′ - 12,000′
Montana to British Columbia, s. to Colorado, Arizona and California

Bracts below flowers usually blackish to dark brown.

47

Western Yarrow

Milfoil

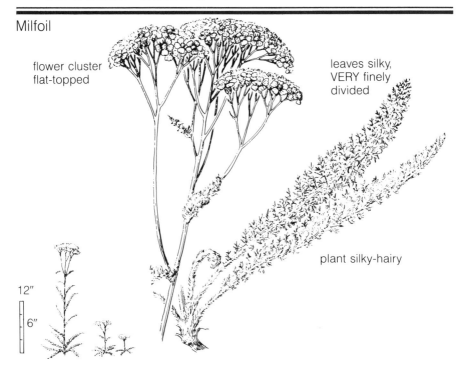

flower cluster
flat-topped

leaves silky,
VERY finely
divided

plant silky-hairy

12"

6"

Achillea lanulosa   Sunflower family

Blooms white. June-September. 5,500' - 11,500'
Manitoba and British Columbia, s. to Kansas, New Mexico, Arizona, California and n. Mexico

Without a doubt, the western yarrow is one of the best known of all mountain flowers. The long, lacy, fernlike leaves are helpful in identifying the plant.

People from the eastern States may be familiar with the common yarrow of that area, *Achillea millefolium,* which looks much the same as this one. The eastern species was introduced from Europe, while ours is native.

Since ancient times the plant has been highly regarded for its healing properties. Legend ascribes the discovery of its virtues to Achilles, in whose honor it is named.

According to the Zuñi, the leaves produce a cooling sensation when applied to the skin. Because of this the flowers and roots are chewed and the mixture is rubbed on the bodies of participants in ceremonies involving the use of fire.

Among the superstitious, yarrow is sometimes regarded as a potent love charm, especially if picked from the grave of a young man by a lovelorn maiden.

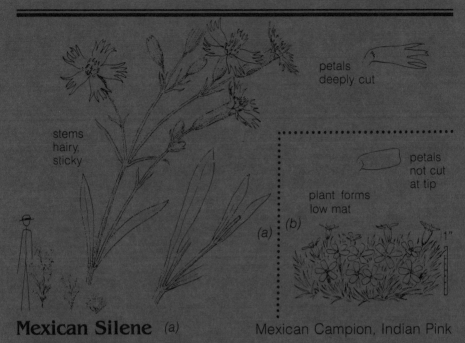

petals
deeply cut

stems
hairy,
sticky

petals
not cut
at tip

plant forms
low mat

(a) (b)

1"

Mexican Silene (a)

Mexican Campion, Indian Pink

Silene laciniata

Pink family

Blooms brilliant red. July-October. 5,500' - 9,000'
W. Texas to California and Mexico

One of the truly handsome flowers of the Southwest mountains is the showy Mexican silene with its scarlet flowers measuring up to an inch and a half in diameter. Each of the five petals is slashed at the tips into four divisions.

Many members of the pink family have their petals divided in this manner; in fact, this characteristic of the petals being "pinked" has given the family its name.

Moss Silene (b)

Cushion Pink

Silene acaulis

Pink family

Blooms purplish-pink. July-September. 11,000' - 12,000'
On high mountains and in cold climates of both the Old and New Worlds

When you are at higher elevations and particularly above timberline, be on the lookout for this small, but attractive member of the pink family. It often grows in cushionlike mats resembling clumps of coarse moss, and is covered in the summer with pretty little flowers.

During the blooming period the flowers form large patches of color among the rocks and tundra above timberline. They last only a short time.

Skyrocket Gilia

Trumpet Phlox

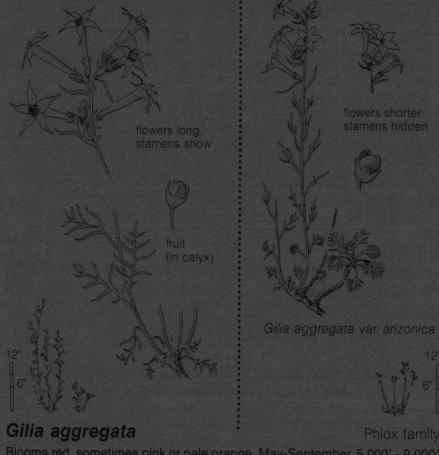

flowers long, stamens show

fruit (in calyx)

flowers shorter, stamens hidden

Gilia aggregata var. *arizonica*

12″
6″

12″
6″

Gilia aggregata

Phlox family

Blooms red, sometimes pink or pale orange. May-September. 5,000′ - 9,000′
Montana to British Columbia, s. to New Mexico, Arizona and California

The brilliant red, trumpet-shaped flowers of the skyrocket gilia make it one of the Southwest's most handsome wild flowers. It is quite common on sunny slopes in the ponderosa pine belt, and is often seen along the roadsides.

The Navajos are reported to use the dried leaves for treating various kinds of stomach trouble. Hopis grind the flowers with meal for the hunter's offering before setting out to hunt pronghorns, which are fond of the plant.

A variety of *Gilia aggregata*, the subspecies *arizonica*, usually called "Arizona gilia," is considered by some botanists to be a separate species. With its smaller flowers, shorter stems, and bushier growth habit, it looks considerably different from the typical *aggregata*.

Penstemons

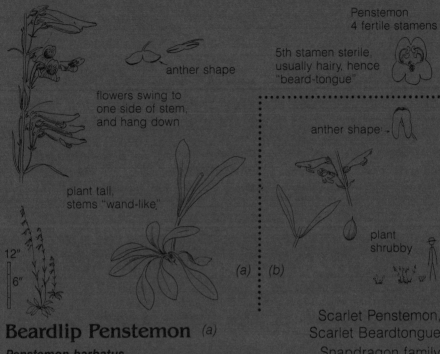

Penstemon
4 fertile stamens

5th stamen sterile,
usually hairy, hence
"beard-tongue"

anther shape

flowers swing to
one side of stem,
and hang down

anther shape

plant tall,
stems "wand-like"

plant
shrubby

12"
6"

(a) (b)

Beardlip Penstemon *(a)*

Penstemon barbatus

Scarlet Penstemon,
Scarlet Beardtongue
Snapdragon family

Blooms red. June-October. 4,000' - 10,000'
S. Colorado and Utah to central Mexico

    This is the common red penstemon so often seen along the roadsides of northern Arizona and New Mexico.

    One variety is used by the Navajos, and sometimes by native New Mexicans, in treating kidney disorders. The latter also make a syrup of the boiled flowers for whooping cough. Zuñi Indians chew the root, and rub it on their rabbitsticks to assure a successful hunt.

Bridges Penstemon *(b)*

Penstemon bridgesii   Snapdragon family

Blooms red. May-September. 4,500' - 7,500'
Sw. Colorado and w. New Mexico to California

    Often associated with beardlip penstemon, but rarely grows higher than the ponderosa pine belt. The generic name is sometimes spelled *Pentstemon*, because the flowers all have five stamens. One of them is flattened, usually hairy, and curved downward, and the name "beard-tongue" is derived from this.

Painted-Cups

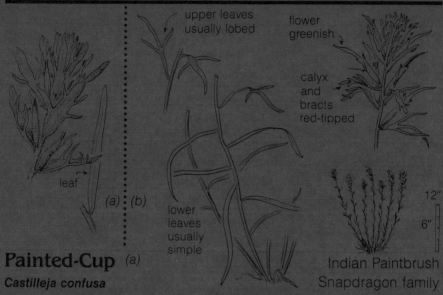

upper leaves usually lobed

flower greenish

calyx and bracts red-tipped

leaf

(a) (b)

lower leaves usually simple

12"

6"

Painted-Cup (a)

Castilleja confusa

Indian Paintbrush
Snapdragon family

Blooms scarlet. June-October. 7,000' - 10,000'
Colorado, n. New Mexico and n. Arizona

Many species of painted-cup are found in the United States, and lots of them are partially parasitic on the roots of other plants. The plants are named after the Spanish botanist, Castillejo, and one of the species is now the state flower of Wyoming.

Wyoming Painted-Cup (b) Narrow-leaved Indian Paintbrush

Castilleja linariaefolia

Snapdragon family

Blooms red. April-October. 5,000' - 10,000'
Wyoming to New Mexico, Arizona, s. Nevada and California

Those who are familiar with the handsome Indian paintbrush may be surprised to find that the true flowers of this plant are small and quite inconspicuous. The large splashes of red which most people consider as flowers are actually brightly colored bracts or leaves, among which the small, usually greenish flowers are almost hidden.

These plants, where abundant, are of some value in providing forage for livestock. The Zuñi Indians use the root of Indian paintbrush in conjunction with various minerals for coloring deerskin black. In the Hopi village of Hano the red flowers are painted on pottery and carved in wood, and girls wear them for adornment.

Spanish New Mexicans use a decoction of the plants for treating kidney disorders; the dried plant, mixed with other ingredients, is used for skin trouble and supposed cases of leprosy.

Fairyslipper

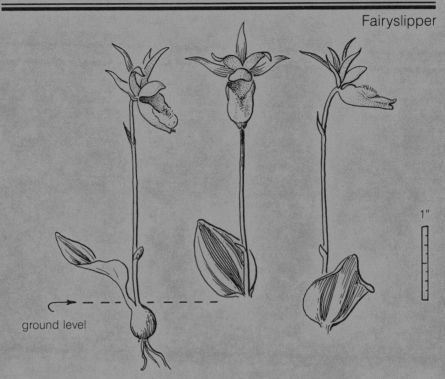

ground level

1"

Calypso bulbosa   Orchid family

Blooms pinkish to rose-purple. June-July. 8,000' - 10,000'
Labrador to Alaska, s. to New England, Michigan, Arizona and California,
also in Europe

Most of the orchids found in North America are rather small and inconspicuous, but the large and showy ladyslipper and the dainty fairyslipper
are two exceptions.

The typical orchid blossoms of the fairyslipper are borne singly on stems
from 4 to 8 inches tall, which grow from a bulbous root. The large pinkish
petal at the bottom of the flower is usually marked and spotted with a darker
color, and it is often bearded with bright yellow hairs. There is but one large,
broad leaf located at the base of the flower stalk.

If you plan to see this dainty beauty in bloom, you will have to head for
the mountains in June or July. Your search may be rewarded if you look
diligently in moist pine and spruce woods and along cool, shady streams.
If luck is with you, please only admire the beauty, and do not pick it. It would
be a pity for this orchid to become as rare as the beautiful yellow ladyslipper.

Indians of the northwestern States and Alaska are reported to have eaten
the small bulbs either raw or cooked.

Stonecrop

Rose Crown

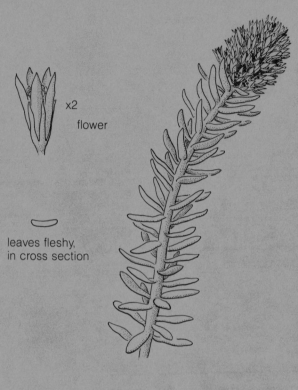

x2
flower

leaves fleshy,
in cross section

12"

6"

Sedum rhodanthum

Orpine family

Blooms pink to white. July-September. 9,000' - 12,000'
Montana to Utah, New Mexico and Arizona

The stiff stems of this succulent plant bear heads or spikes of rose pink to white flowers. Stems range from 4 to 15 inches in height. This plant should be looked for along streams and other moist places from the Canadian through the Arctic-Alpine Life Zone.

Many plants in the orpine family are cultivated as ornamentals, and are particularly popular with succulent plant fanciers. Various species of *Sedum* are grown extensively in rock gardens.

Stonecrop

King's Crown, Roseroot

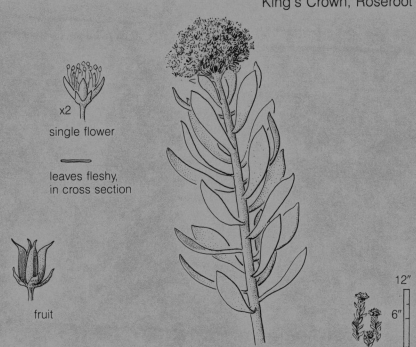

x2
single flower

leaves fleshy,
in cross section

fruit

12"

6"

Sedum integrifolium   Orpine family

Blooms deep red to purple. July-September. 9,000' - 14,000'
Montana to New Mexico, Utah and possibly Arizona

In contrast to the flowers of the plant described above, the tiny, deep-red flowers of "king's crown" grow in rounded clumps at the ends of un-branched stems from 4 to 12 inches high. Note that the flowers of "rose crown" come out from the sides of the stems. All of the members of the genus *Sedum* have smooth fleshy leaves which aid in storing water for the plant.

In Greenland, it is reported that the plant is used for a salad and poultices for headaches are made from the leaves. A species of *Sedum* is used by native New Mexicans for the treatment of various disorders. For earache one of the leaves is heated and placed in the ear, or if someone should suffer from corns, a leaf is crushed and bound on them. Next time you wake up with a backache, try an old New Mexican remedy which consists of mashing a handful of *Sedum* leaves, heating them, and applying them as a poultice to the back.

The name "roseroot," which is sometimes used for this species, is due to the pleasant rose-like odor of the root.

Alumroot

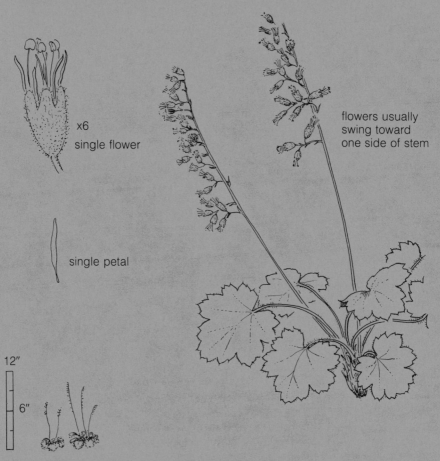

x6
single flower

flowers usually
swing toward
one side of stem

single petal

12"

6"

Heuchera versicolor   Saxifrage family

Blooms pinkish. June-October. 6,500' - 12,000'
W. Texas and s. Utah to Arizona, California and n. Mexico

This species of alumroot is an attractive plant and would be suitable for cultivation in rock gardens wherever the climate is cool enough. The pink flowers are borne on slender, leafless stems which spring from a somewhat woody base surrounded by leaves.

You can usually find this plant in coniferous forests, especially around moist and shaded rocks.

As indicated by the common name, the rootstalks of these plants have astringent properties. They have been used by woodsmen and others in cases of diarrhea.

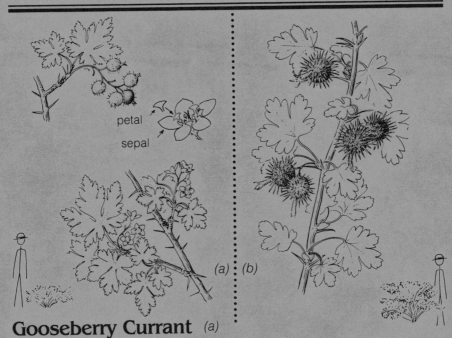

petal

sepal

(a) (b)

Gooseberry Currant (a)

Ribes montigenum

Saxifrage family

Blooms pink to red. June-August. 6,500' - 11,500'
Montana to British Columbia, s. to New Mexico, Arizona and California

Both gooseberries and currants are members of the genus *Ribes*. Usually they may be distinguished by the stems which have spines and prickles in the gooseberries and are more or less smooth in the currants. However, the gooseberry currant is an exception to this rule for it is a currant with spiny stems. This doubtless accounts for the common name.

The edible berries are bright red and are crowned with the withered remains of the flower. Both domestic animals and deer browse the plant, and the berries are much liked by birds.

Orange Gooseberry (b)

Ribes pinetorum

Saxifrage family

Blooms reddish. April-September. 7,000' - 10,000'
New Mexico and Arizona

In the mountains of southern Arizona this is the most abundant and the handsomest of the wild gooseberries. It has reddish flowers and large, densely prickly berries, which turn purple at maturity.

These plants harbor a stage of the white pine blister rust, and are being destroyed wherever white pines occur in commercially important stands.

Roses

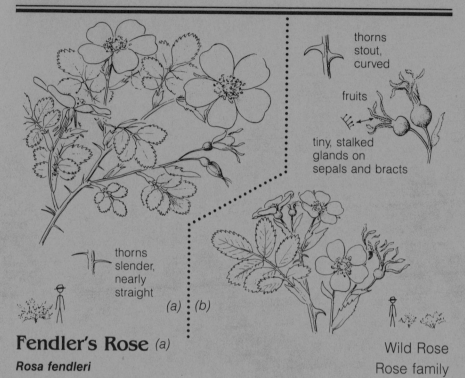

thorns stout, curved

fruits

tiny, stalked glands on sepals and bracts

thorns slender, nearly straight

(a) : *(b)*

Fendler's Rose *(a)*

Rosa fendleri

Blooms pink. July-August. 5,500' - 9,000'
Minnesota to British Columbia, s. to n. Mexico

Wild Rose
Rose family

    Almost everyone will recognize the wild roses from their decided resemblance to the cultivated varieties. This plant may reach a height of 4 or 5 feet. The pink flowers are about 2 inches in diameter.

    The Navajos eat the red fruits, and the wooden needles they use in leatherwork are sometimes made of the rose bush. Some Spanish New Mexicans use dried and ground petals for treating sore throats and healing sores.

Arizona Rose *(b)*

Rosa arizonica

Rose family

Blooms pink. May-July. 4,000' - 9,000'
New Mexico and Arizona

    This is the most widespread and abundant wild rose in Arizona. Look for it in the shade along streams in pine forests.

    The fruits are eaten by birds and other animals and Hopi Indian children are reported to eat them occasionally. Where they form thickets, which is unusual in the Southwest, wild roses are valuable in controlling soil erosion and provide nesting and protective cover for wildlife.

Bearberry

Kinnikinnick, Manzanita

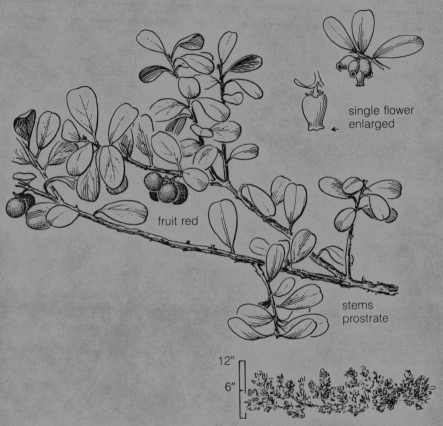

single flower
enlarged

fruit red

stems
prostrate

12"
6"

Arctostaphylos uva-ursi

Heather family

Blooms pinkish-white. May-June. 8,000' - 10,000'
Most of the cooler parts of the northern hemisphere

Low, creeping, evergreen shrubs rarely more than 2 feet high. In common with other manzanitas the bark is reddish, leaves are thick and leathery, and the waxy flowers are small and bell-shaped.

Bearberry is one of Nature's pioneers, as it is one of the first plants to grow on bare and rocky ground and on burned over areas. Apparently the plant is not too appetizing, for goats are the only animals to consistently browse it, although deer make some use of the foliage. The round, red berries are relished by many animals, especially by bears.

Preparations made from bearberry are sometimes used for treating urinary disorders, rheumatism, and anemia.

Greenleaf Manzanita

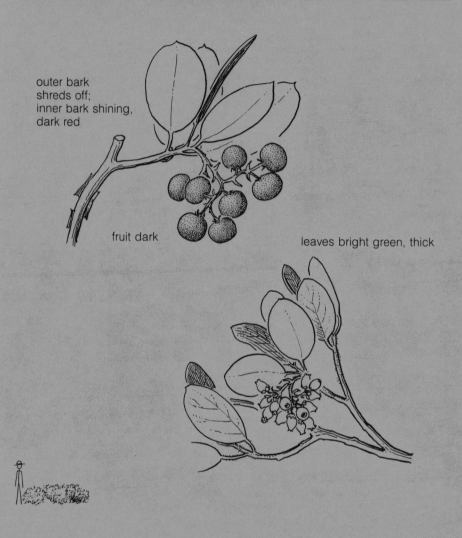

outer bark
shreds off;
inner bark shining,
dark red

fruit dark

leaves bright green, thick

Arctostaphylos patula

Heather family

Blooms pinkish-white. May-June. 7,000' - 8,500'
Colorado to n. Arizona and California

Usually not more than 4 feet high, this species of manzanita is distinguished by its leaves, which are almost round and are a bright green. The plants form thickets and the stems root where they touch the ground.

Greenleaf manzanita is highly fire resistant.

Spreading Dogbane

Indian Hemp

flowers,
enlarged

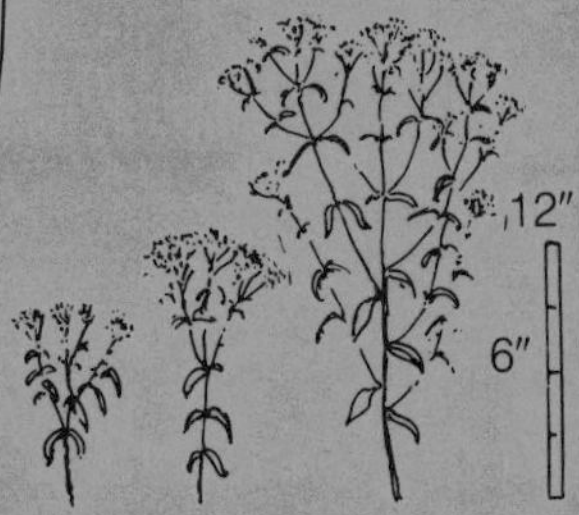

12"

6"

Apocynum androsaemifolium

Dogbane family

Blooms pink to white. June-July. 7,000' - 9,000'
Canada southward to Georgia and Arizona

The dogbanes are so called because the Greek name alludes to an old belief that these plants were poisonous to dogs. Most of the members of this family grow in the tropics and several are definitely poisonous. Most, if not all, of our species are poisonous to livestock because of the presence of a toxic substance known as cyanogen.

Early American Indians used the bark of *Apocynum* for making cordage. When broken, the stems exude a milky sap, which, in the case of *Apocynum cannabinum,* is dried and used as chewing gum in some parts of New Mexico.

The oval leaves are dark green above, but paler or whitish below.

Field Mint

Wild Mint

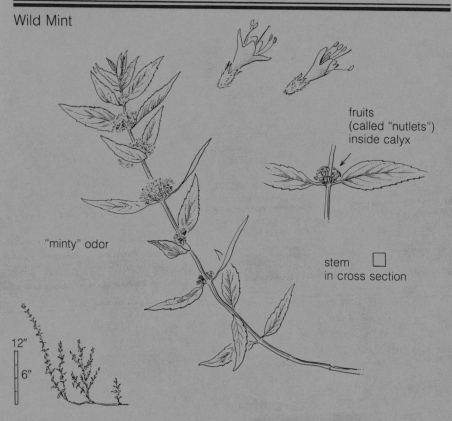

fruits
(called "nutlets")
inside calyx

"minty" odor

stem ☐
in cross section

12"
6"

Mentha arvensis

Mint family

Blooms pink. July-October. 5,000' - 9,500'
N. North America, Europe and Asia

Members of the mint family may be recognized by their square stems, opposite leaves, and usually aromatic foliage. The odor of mint is strong when the leaves of this plant are crushed. At the bases of the upper leaves there are dense clusters of small, pink flowers.

Everyone is familiar with the use of mint for flavoring, but the medicinal uses are not so well known. The Navajos apply a preparation of the plant for the treatment of swellings. Oil of peppermint, extracted from the Old World species, *Mentha piperita,* is used in medicine as well as for flavoring. The leaves of this mint are used in various ways by some Spanish New Mexicans for treating fever. It is said that mint tea is an excellent stomach tonic.

Spearmint, *Mentha spicata,* is a species native to the Old World, but is now widely distributed in America. It may be distinguished from the field mint by its flowers, which are in spikes at the top of the flower stalk.

Snowberries

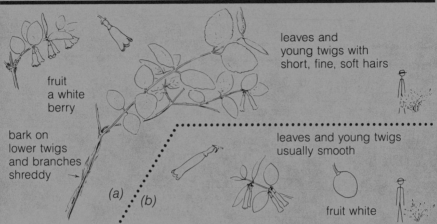

fruit
a white
berry

leaves and
young twigs with
short, fine, soft hairs

bark on
lower twigs
and branches
shreddy →

(a)   (b)

leaves and young twigs
usually smooth

fruit white

Roundleaf Snowberry (a)

Symphoricarpos rotundifolius

Honeysuckle family

Blooms pink to white. May-June. 4,000' - 10,000'
S. Colorado, New Mexico and Arizona

As a group, the snowberries are branching shrubs with shreddy bark and clusters of small, bell-shaped flowers. In the fall they bear attractive waxy white berries which are relished by birds.

The roundleaf snowberry may be distinguished from the following species by the dense fuzz or hair on the young twigs, and the soft fuzziness on both surfaces of the leaves. Slender hairs are usually found inside the flower itself.

All snowberry fruits are important wildlife foods in the western states. They are heavily used by the sharp-tailed and Franklin grouse, pine grosbeak, robin and varied thrush. Deer and pronghorns eat the foliage.

Mountain Snowberry (b)

Symphoricarpos oreophilus

Honeysuckle family

Blooms pink to white. May-August. 5,500' - 9,000'
Colorado and w. Texas to e. Nevada, Arizona and n. Sonora

A shrub about 3 feet high with small tubular flowers about half an inch long arranged in pairs. This species lacks the fuzziness of the roundleaf snowberry.

The plants contain small quantities of saponin, but cases of poisoning are rarely reported and the plants are extensively browsed by deer and livestock.

Some species make excellent borders for shrubbery and for covering the ground under trees. They spread by suckers, and will thrive in almost any soil. Propagation is effected by cuttings, by division, and also by seeds.

Valerians

Sharpleaf Valerian (a)
Valeriana capitata **ssp.** *acutiloba*

Tobacco-root
Valerian family

Blooms pink to white. May-July. 7,000' - 9,500'
Wyoming to New Mexico and Arizona

Your chances of finding this plant are best in moist mountain meadows. The leaves are mostly at the base of the plant and they are usually shaped like a spatula.

Some species of valerian are used in medicine as a sedative and for nervous troubles. The root has a particularly unpleasant smell which lasts for years after the plant is dried.

Indians of the northwestern states and Canada are said to have cooked the roots in stone-lined pits, and also to have used the seeds for food. The leaves are supposedly efficacious in healing wounds.

Arizona Valerian (b)
Valeriana arizonica

Valerian family

Blooms pink to purplish-pink. May-June. 4,500' - 8,000'
S. Utah to s. Arizona

Not found at the higher elevations, but locally common in the ponderosa pine belt of Arizona. Often growing in rock crevices where moisture is present. An attractive plant from 3 to 10 inches tall, having smooth hollow stems.

Wyoming painted-cup, *Castilleja linariaefolia,* and goldenpea, *Thermopsis pinetorum*

Field mint, Mentha arvensis

Indian paintbrush, *Castilleja confusa*   Skyrocket gilia, *Gilia aggregata*

Beardlip penstemon, *Penstemon barbatus*

Beardlip penstemon, and false solomonseal, *Smilacina racemosa*

Tall false dandelion, *Agoseris glauca*

Ladyslipper

12"

6"

Cypripedium calceolus

Orchid family

Blooms yellow. June-July. 6,000' - 9,000'
Newfoundland to British Columbia, s. to Georgia, Arizona and Washington

One of the rarest and most beautiful flowers of the Southwest mountains is this handsome ladyslipper. The chances of finding it are slight, but if ever you do, you will want to know its name, and because of this it has been included in the booklet.

The flowers of all orchids are irregularly shaped and are composed of six main parts. There are three sepals colored like petals and three petals, one of which is much larger than the others and is called the lip. In the ladyslipper this petal is yellow and shaped like a shoe; hence the common name. The other petals and sepals are yellowish-green with dull reddish markings. There is usually but one flower on a stem from 8 to 20 inches tall.

If you should be so fortunate as to find one of these beautiful flowers, please don't pick it, as the plant is in danger of extermination.

Eriogonums

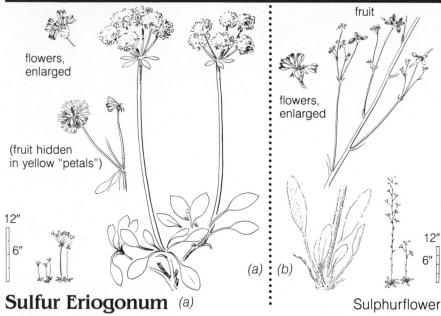

flowers,
enlarged

(fruit hidden
in yellow "petals")

12"

6"

(a) : *(b)*

fruit

flowers,
enlarged

12"

6"

Sulfur Eriogonum *(a)*

Eriogonum umbellatum

Sulphurflower

Buckwheat family

Blooms yellow. April-September. 5,000' - 9,000'
Wyoming to Washington, n. Arizona and California

Easily recognized by its umbrella-like clusters of bright, yellow flowers. The spatula-shaped leaves are white and fuzzy underneath, but smooth and green on top. Late in the season the flowers may turn reddish.

The scientific name, *Eriogonum,* is from the Greek, meaning woolly knees, and refers to the woolly joints on the stems of many species.

More than 200 species are known in this North American genus. The seeds and other plant parts are of moderate importance as food for wildlife.

Wing Eriogonum *(b)*

Eriogonum alatum

Winged Buckwheat

Buckwheat family

Blooms yellow. July-September. 5,500' - 9,500'
Nebraska to Texas, Utah and Arizona

A tall, branched plant with small, yellow flowers followed by winged, triangular seeds. The leaves are from 1 to 3 inches long, fuzzy on the upper surface, and grow mostly at the base of the plant.

The Navajos eat the roots raw or dried, and they are used medicinally to alleviate pain. The powdered root is used in much the same manner by the Zuñi, who take the root in warm water to relieve a "general miserable feeling."

Golden Columbine

Yellow Columbine

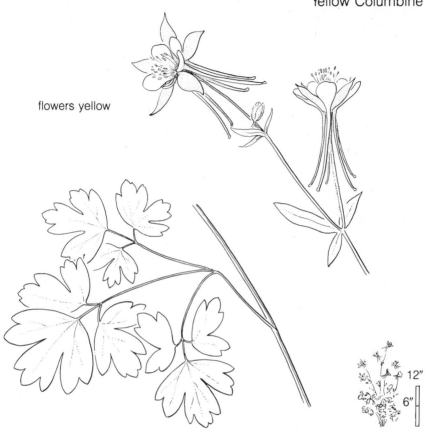

flowers yellow

12"
6"

Aquilegia chrysantha   Buttercup family

Blooms yellow. April-September. 3,000' - 11,000'
S. Colorado to New Mexico, Arizona and n. Mexico

One of the handsomest flowers of the Southwest is this columbine with its large, long-spurred, canary-yellow flowers. In Arizona this is the most common species of columbine, growing abundantly in the pine belt.

This plant is exceptional in its wide altitudinal range. Under favorable conditions of moisture and temperature it may grow as low as 3,000 feet, but it is more often found in rich soil at higher elevations up to 11,000 feet or more.

Columbines make attractive plants for the flower garden, but the different species should be grown as far apart as possible, for they readily hybridize and the resulting seed is often impure.

Buttercups

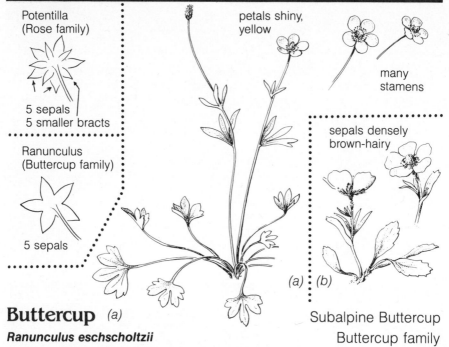

Potentilla
(Rose family)

5 sepals
5 smaller bracts

Ranunculus
(Buttercup family)

5 sepals

petals shiny,
yellow

many
stamens

sepals densely
brown-hairy

(a)   (b)

Buttercup (a)

Ranunculus eschscholtzii

Subalpine Buttercup
Buttercup family

Blooms yellow. July-August. 10,000' - 12,000'
Alaska to Colorado, n. Arizona and California

    Many species of the familiar buttercup are found in the mountains of the Southwest. This one is shiny and smooth and usually grows in moist places near timberline.

    Buttercups are easy to recognize due to their shiny yellow petals that look as though they had been varnished, and their usually much-divided leaves. They are sometimes confused with the cinquefoils, belonging to the rose family, but a glance at the small drawing (in the margin) will show you how to distinguish between them.

    You have probably heard the old story that the rich color of butter in early spring is due to cattle eating buttercups. This is rather doubtful, for most buttercups have a bitter juice, unpalatable to cattle and other animals.

Macauley Buttercup (b)

Ranunculus macauleyi

Alpine Buttercup
Buttercup family

Blooms yellow. June-July. 10,000' and above
Colorado and n. New Mexico

    Handsome plant in Arctic-Alpine meadows, often growing at the very edge of melting snowbanks.

Creeping Barberry

Oregon Grape

12"

6"

Berberis repens

Barberry family

Blooms yellow. April-June. 5,000' - 8,500'
British Columbia to Wyoming, New Mexico, Arizona and California

Because of its spiny-edged leaves, this low shrub is often mistaken for holly. It is often found on dry, rocky slopes where its creeping rootstalks provide excellent ground cover and aid in preventing erosion.

The berries make a delicious jelly, especially when combined with apple or some other fruit. They are also much used as food by wildlife. From the roots a brilliant yellow dye may be obtained, and a substance in the plant, called berberin, has limited use as a drug.

Spanish New Mexicans boil the leaves or roots in water and drink the tea to cure anemia and cleanse the blood. The Navajos are said to use a decoction of the leaves and branches in the treatment of rheumatism.

The plant is colorful throughout almost the entire year. The small, yellow flowers give a dash of color in the spring, and later in the summer the large, bluish-purple berries hang in small clusters. The colder temperatures of fall bring out shades of red, yellow, and purple in the tough, spiny leaves.

Drabas

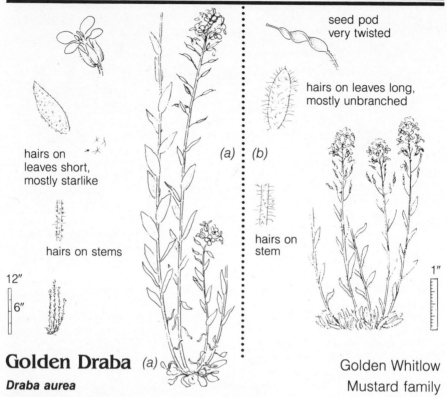

hairs on leaves short, mostly starlike

hairs on stems

(a) : *(b)*

12″
6″

seed pod very twisted

hairs on leaves long, mostly unbranched

hairs on stem

1″

Golden Draba *(a)*

Draba aurea

Golden Whitlow

Mustard family

Blooms yellow. July-August. 5,000′ - 12,000′
Throughout the cooler parts of the northern hemisphere

In the spring and summer this little plant, bearing small, golden-yellow blossoms, may be found growing in the spruce and alpine belts. It favors sunny, open spots in aspen groves and coniferous forests.

The leaves, stems, and often even seed pods are covered with fuzzy hair. Leaves are thick and pods are somewhat twisted.

Draba *(b)*

Draba streptocarpa

Twisted-pod Draba

Mustard family

Blooms yellow. May-July. 9,000′ - 13,000′
Grows in Wyoming, New Mexico and Utah

This species of draba is a free-blooming plant with golden-yellow flowers. The pods, which are characteristic of plants belonging to the mustard family, are smooth and quite twisted.

On the higher mountains above timberline the plant may be small due to unfavorable growing conditions.

Coast Erysimum

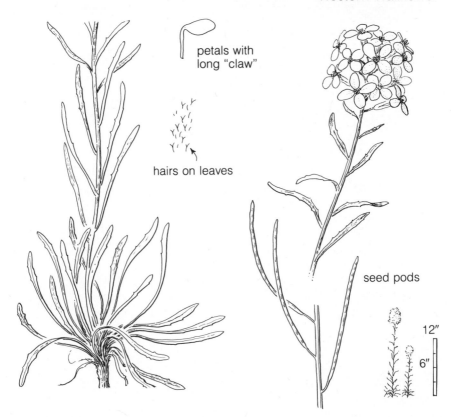

petals with
long "claw"

hairs on leaves

seed pods

12"

6"

Erysimum capitatum   Mustard family

Blooms yellow to dark orange or maroon. March-September. 2,500' - 12,000'
Saskatchewan to Washington, s. to New Mexico, Arizona and California

A common plant occurring in a wide altitudinal range. Plants of higher elevations may vary from typical *Erysimum capitatum* by having orange to maroon flowers instead of bright yellow. Some botanists regard this high altitude form as a distinct species and call it *Erysimum wheeleri.*

When in flower the plants are really quite handsome; and certainly the common name, wallflower, is a misnomer. However, in Europe the plants often grow against old walls, hence the name. The true wallflower belongs to a different genus, *Cheiranthus.*

Zuñi Indians grind the entire plant and mix it with water. The resulting mixture is then applied to the forehead and temples to relieve pain caused by overexposure to heat, and as a preventative against sunburn.

75

Saxifrages

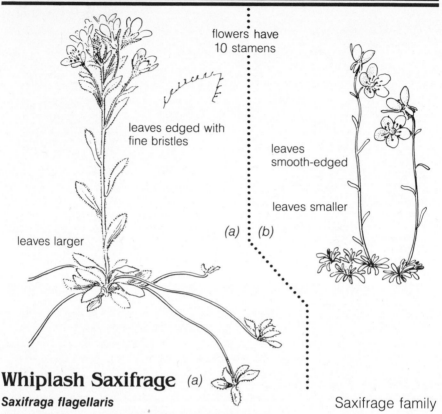

flowers have
10 stamens

leaves edged with
fine bristles

leaves
smooth-edged

leaves smaller

(a) (b)

leaves larger

Whiplash Saxifrage (a)

Saxifraga flagellaris   Saxifrage family

Blooms yellow. July-September. 10,000' - 14,000'
In the far north and south in the Rockies to New Mexico and n. Arizona

Most members of this genus grow only on high mountains, often among the rocks above timberline. This one is a small plant from 1 to 8 inches high bearing one or very few bright yellow flowers. It gets its name from the presence of whip-like runners growing out from the base.

Goldbloom Saxifrage (b)

Saxifraga chrysantha   Saxifrage family

Blooms yellow. July-September. 10,000' - 14,000'
Colorado and n. New Mexico

A dainty plant growing only 1 to 3 inches high from a basal cluster of small leaves. It is rare indeed to find it anywhere but among the rocks in the Arctic-Alpine Life Zone. The yellowish petals have orange spots and the yellow seed pod turns red upon ripening. The plant often grows in clumps in the shelter of a protruding rock.

Bush Cinquefoil

Shrubby Cinquefoil

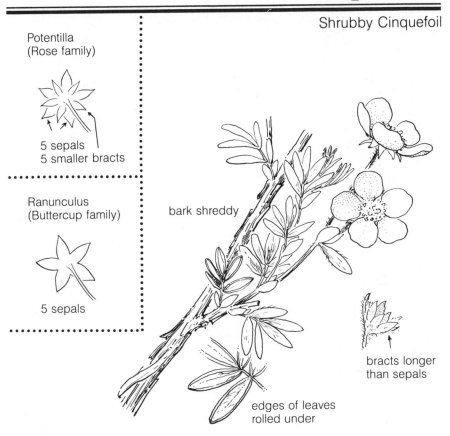

Potentilla
(Rose family)

5 sepals
5 smaller bracts

Ranunculus
(Buttercup family)

5 sepals

bark shreddy

bracts longer
than sepals

edges of leaves
rolled under

a leafy shrub

Potentilla fruticosa

Rose family

Blooms yellow. June-August. 7,000' - 9,500'
Cooler parts of the northern hemisphere

This is the only shrubby species of *Potentilla* in the Southwest. It is a handsome shrub when in bloom and may be found along streams and in moist meadows.

Heavy browsing by sheep, goats, and deer sometimes stunt the plants. Bush cinquefoil is excellent for the control of soil erosion.

Cinquefoils

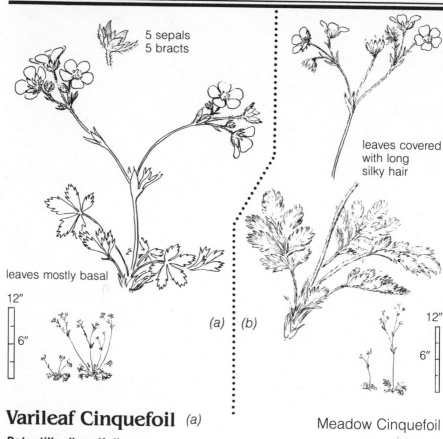

5 sepals
5 bracts

leaves covered
with long
silky hair

leaves mostly basal

12"

6"

(a) *(b)*

12"

6"

Varileaf Cinquefoil *(a)*

Potentilla diversifolia

Blooms yellow. July-September. 8,000' - 12,000'
W. Canada to New Mexico, Arizona and California

    The name cinquefoil comes from the French and means five leaves. These plants are also called five fingers because the leaves of some species, with their five leaflets, simulate a hand with the fingers spread. This is well illustrated by the varileaf cinquefoil.

Meadow Cinquefoil

Rose family

Horse Cinquefoil *(b)*

Potentilla hippiana

Blooms yellow. June-September. 7,000' - 11,500'
Pea family. Blooms yellow. 6,000' - 9,500'

    A powder prepared from this plant is used by the Navajos for healing sores. It is also used in treating burns and to expedite childbirth.

Silver-leaf Cinquefoil

Rose family

78

Pine Thermopsis

Goldenpea

fruit hairy

cross section
of pods flat

stamens separate
and distinct
(inside keel petal)

calyx-teeth
equal

Thermopsis pinetorum

Pea family

Blooms yellow. 6,000' - 9,500'
Colorado, Utah, New Mexico and Arizona

A handsome plant with large, bright yellow flowers growing mostly in pine forests. Due to its habit of spreading by rootstalks, the plants are often found clumped.

The pea-shaped flowers readily identify the family to which this plant belongs. After the flowers disappear, long, flat pods with several seeds develop. Under favorable conditions the flower stems may reach a height of 2 feet or more.

Some species of *Thermopsis* are thought to be poisonous to stock. Several cases of poisoning of children who have eaten the seeds have been reported.

About 18 species of *Thermopsis* are known from North America and northern and eastern Asia. They do best in deep, light, well-drained soil. Because of their deep-rooted nature they are usually able to withstand drought very well.

Southwestern St. Johnswort

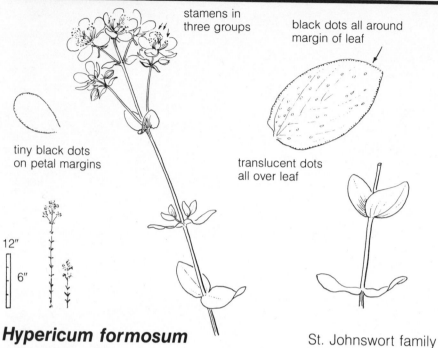

stamens in
three groups

black dots all around
margin of leaf

tiny black dots
on petal margins

translucent dots
all over leaf

12"

6"

Hypericum formosum

St. Johnswort family

Blooms yellow with black dots. July-September. 5,000' - 9,000'
Wyoming to Arizona, s. California and Mexico

Certainly this pretty little plant deserves a more attractive name. The name, St. Johnswort, stems from the fact that they bloom about St. John's Day.

The dainty flowers are borne at the top of a stiff stem, often branching at the top, from 6 inches to 3 feet in height. The dull green leaves clasp the stem in pairs. Usually there are many black, or occasionally translucent, dots along the edges of the leaves.

While not abundant, the plant occurs widely through the Southwest, and is often found in moist mountain meadows.

A European species, *Hypericum perforatum,* widely distributed in the United States, contains a substance that may cause sensitiveness to sunlight. White-skinned horses, cattle, and sheep, eating the plant and then exposed to bright sun, sometimes blister and suffer loss of hair. It has not been proven that our native species have this same effect.

Some species of St. Johnswort are cultivated as ornamentals in gardens. They are said to thrive in good, loamy soil and even in sandy soil, if sufficiently moist. They should be grown in partly shaded situations and will bloom longer if not exposed to the full sun. As a rule they are short-lived plants.

Hooker Evening-Primrose

fruit

Oenothera hookeri

Evening-primrose family

Blooms yellow, fading to pink. July-October. 3,500' - 9,500'
British Columbia to Montana, s. to Mexico

A handsome plant from 2 to 6 feet in height bearing striking, yellow flowers as much as 3 inches across. As the shadows of early evening slant across the mountain meadows, the blossoms commence to open, but by the noonday sun of the following day their beauty has withered away. The withered blossom is usually a pinkish color.

The yellow pollen is connected by filament-like threads which cling to visiting insects such as the pink night moth. Thus it is carried from flower to flower.

Evening-primroses were formerly cultivated in England for their edible roots. Coughs, colds and asthmatic troubles have been treated with a drug made from the evening-primrose.

Some of our most prized cultivated flowers are members of this family. Representatives of such genera as *Fuchsia, Clarkia,* and *Godetia* are found in many flower gardens.

Pseudocymopterus

Mountain Parsley

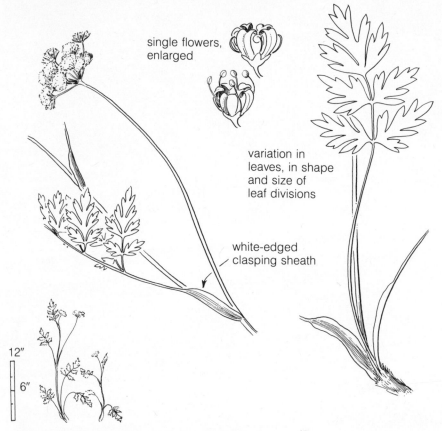

single flowers, enlarged

variation in leaves, in shape and size of leaf divisions

white-edged clasping sheath

12"

6"

Pseudocymopterus montanus　　　　Parsley family

Blooms variable, yellow to purple. May-October. 5,500' - 12,000'
S. Wyoming to w. Utah, s. to n. Mexico

As is true of most members of the parsley family, the small yellow flowers of this plant are grouped in a flat-topped cluster at the top of the stem. The plants grow from 1 to 2 feet tall and are usually fairly common in pine woods, although they may extend above timberline.

The leaf shapes and flower colors are highly variable and seem to be controlled to a great extent by the habitat of the individual plant. Flowers on the same plant may sometimes vary from yellow through orange — even to purple. However, yellow is the predominant color.

"Mountain parsley" is a relative of the parsley of kitchen gardens and of sweet anise, used to flavor confectionery.

82

Manyflower Gromwell

Manyflower Puccoon

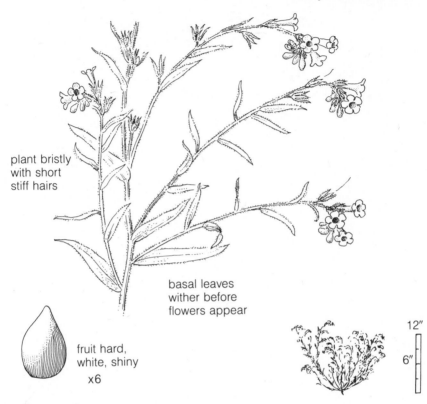

plant bristly
with short
stiff hairs

basal leaves
wither before
flowers appear

fruit hard,
white, shiny
x6

12″

6″

Lithospermum multiflorum   Borage family

Blooms orange-yellow. June-September. 6,000′ - 9,500′
Wyoming to New Mexico and Arizona

Plants from 1 to 3 feet high with bright yellow or orange funnel-shaped flowers bending in clusters from the tips of the stems. Because of their hard, white, and shining seeds the genus has been given a Greek name meaning stone seed. The plant is called *plante aux perles* by the French because of the fancied resemblance the seeds to pearls.

This handsome plant is often quite common in gravelly soil of the Transition Zone and slightly lower. The narrow leaves are grayish and both stems and leaves are densely hairy.

The members of the borage family are of little economic importance but several species are widely cultivated as ornamentals. The garden heliotropes and forget-me-nots are examples. The roots of the manyflower gromwell were used by Indians in making purple dye.

Flannel Mullein

Common Mullein

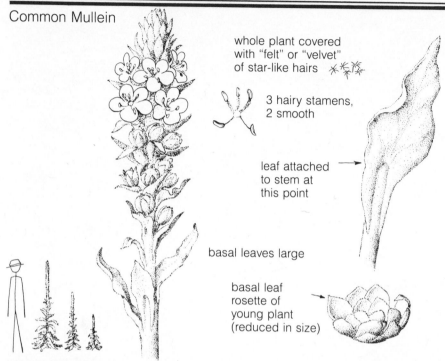

whole plant covered
with "felt" or "velvet"
of star-like hairs

3 hairy stamens,
2 smooth

leaf attached
to stem at
this point

basal leaves large

basal leaf
rosette of
young plant
(reduced in size)

Verbascum thapsus

Snapdragon family

Blooms yellow. Summer. 5,000' - 8,000'
Throughout most of North America

Surely everyone has seen this plant at one time or another. Although it is not a true mountain plant, it is extremely common along roadsides and in waste ground in the ponderosa pine belt. The plant, which was introduced from Europe, is a very prolific seed producer and is now a permanent part of our flora.

The leaves are very distinctive, as they are so thickly covered with woolly hairs that they have the appearance and feel of felt.

The Hopis are said to dry and smoke the leaves and this is thought to cure people who are mentally unbalanced. Early Greeks and Romans dipped dried mullein stalks in tallow to make lampwicks. It may be due to this that the Spaniards named it candelaria, and the English, torchweed.

Early Spaniards used dried mullein leaves as a substitute for tobacco, the crumbled leaves being wrapped in corn husk to make a cigarette. Some people living in isolated villages in New Mexico maintain that mullein cigarettes, besides being pleasurable, are helpful in treating asthma when the smoke is inhaled. External applications of mullein were used by early Californians in treating sprains and pulmonary diseases.

Monkeyflowers

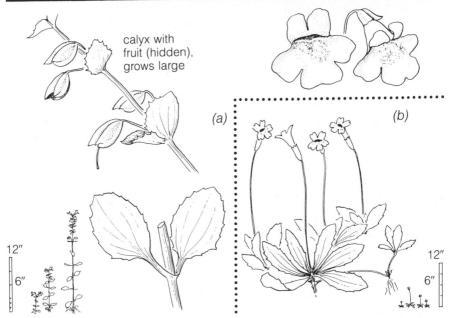

calyx with fruit (hidden), grows large

(a)

(b)

12"

6"

12"

6"

Common Monkeyflower (a)
Yellow Monkeyflower

Mimulus guttatus
Snapdragon family

Blooms yellow, often with reddish spots. March-September. 500' - 9,500'
Montana to Alaska, s. to n. Mexico

Your chances of finding the beautiful yellow monkeyflower are very good if you search in wet meadows, marshy places, and along mountain brooks.

The two-lipped flowers are about 1 inch long and grow on stems from 6 inches to 2 feet high.

The weak stems are rather succulent and have a tendency to develop roots at the joints. The plant can be used for salad and greens.

Primrose Monkeyflower (b)
Little Yellow Monkeyflower

Mimulus primuloides
Snapdragon family

Blooms yellow, sometimes spotted with reddish brown. July-August. 8,000' - 10,000'
Idaho to Arizona and s. California

A dainty little plant from 3 to 6 inches tall bearing single flowers from one-half to 1 inch in length. The toothed leaves are bright yellowish green. It is often found in wet mountain meadows or in marshy areas about springs. The primrose monkeyflower also produces roots at the stem joints, and it tends to form mats.

Honeysuckles

fruit black,
bracts around
fruit turning reddish

edges of
leaves hairy

Bearberry Honeysuckle

Lonicera involucrata

Black Twinberry
Honeysuckle family

Blooms yellow. June-July. 7,500' - 10,500'
Canada and Alaska to Michigan, Arizona, California and n. Mexico

Widely distributed throughout the country the bearberry honeysuckle may often be found growing along streams in open coniferous forests. It is a bush from 3 to 7 feet high which often suckers and tends to form thickets.

The yellow, tubular flowers always occur in pairs and are borne on short stems. There appear a pair of nearly black berries as large as peas.

Birds and small mammals relish the berries, and it is reported that the Indians of British Columbia and Alaska ate the fruit fresh or dried.

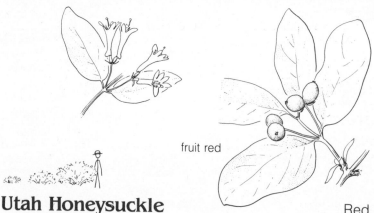

fruit red

Utah Honeysuckle

Lonicera utahensis

Red Twinberry
Honeysuckle family

Blooms yellowish-white. June-July. 9,500' - 11,000'
British Columbia to Montana, s. to New Mexico, Arizona and n. California

Distinguished from the bearberry honeysuckle by the mature fruits which are orange yellow to bright red. The large, oval leaves are whitish beneath and the paired flowers are about three-fourths of an inch long.

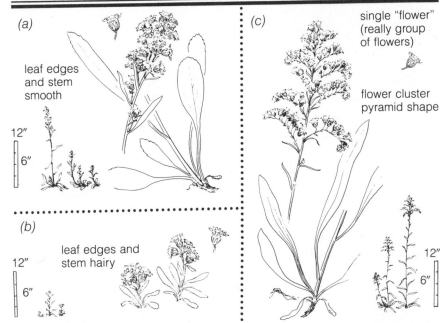

(a)

leaf edges
and stem
smooth

12″
6″

(c)

single "flower"
(really group
of flowers)

flower cluster
pyramid shape

(b)

leaf edges and
stem hairy

12″
6″

12″
6″

Decumbent Goldenrod *(a)*

Solidago decumbens

Dwarf Goldenrod
Sunflower family

Blooms yellow. July-August. 8,000′ - 10,000′
British Columbia to Oregon and Arizona

The most common species of the high Rockies in Colorado. At very high altitudes, its stems may be no more than 1 to 3 inches high, although under more favorable conditions the plant may reach a foot or more.

Goldenrod *(b)*

Solidago multiradiata

Mountain Goldenrod
Sunflower family

Blooms yellow. July-September. 8,500′ - 12,000′
Alberta and British Columbia to n. Arizona and California

Closely resembles the preceding species except the leaves have hairs or bristles along their edges, especially toward the base of the leaf.

Missouri Goldenrod *(c)*

Solidago missouriensis

Prairie Goldenrod
Sunflower family

Blooms yellow. June-August. 5,000′ - 9,000′
Michigan and Tennessee to British Columbia, Oregon and Arizona

Indians of northern Arizona are reported to eat the leaves as a salad.

Cutleaf Coneflower

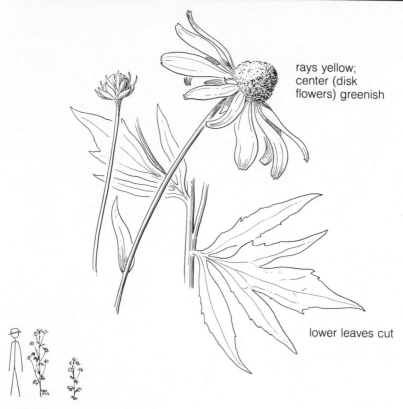

rays yellow;
center (disk
flowers) greenish

lower leaves cut

Rudbeckia laciniata
Sunflower family

Blooms yellow with greenish-yellow center. July-September. 5,000' - 8,500'
Maine to Florida and w. to Idaho, Colorado and s. Arizona

The cutleaf coneflower is a common and showy flower of the Transition
Life Zone where it is often found along brooks and streams. Plants are 3 to
6 feet tall, bearing several-lobed lower leaves and upper leaves usually divided into three parts.

Its yellow sunflower-like flowers are from 3 to 5 inches in diameter and
have drooping, yellow, petal-like rays.

This plant is reported to be poisonous to cattle, sheep, and pigs. The
young stems were once used for food by Indians of New Mexico. A double
form of the flower, known as goldenglow, is cultivated as an ornamental.

A closely related species is the familiar black-eyed-susan with a red-
dish-brown or black center and orange-yellow rays. Although common in
other parts of the country, the black-eyed-susan is not listed for the flora of
Arizona and New Mexico.

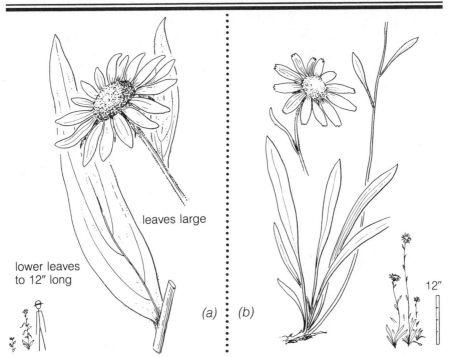

leaves large

lower leaves
to 12" long

(a)   *(b)*

12"

Five-Nerve Helianthella *(a)*

Helianthella quinquenervis

Aspen Sunflower
Sunflower family

Blooms pale yellow. July-October. 5,000' - 10,000'
South Dakota s. to New Mexico, Arizona and Chihuahua

Bears a strong resemblance to the common sunflower, but may be distinguished by its flat seeds and habit of growing at higher elevations. Although the scientific name means little sunflower, this plant reaches a height of 2 to 5 feet. It has thin, five-veined leaves from 4 to 10 inches long, and there are one to several flower heads on each plant from 3 to 6 inches across. This plant is often found in damp aspen groves, moist mountain meadows, and woods.

Parry Helianthella *(b)*

Helianthella parryi

Dwarf Sunflower
Sunflower family

Blooms yellow. July-September. 7,000' - 9,000'
Wyoming to New Mexico and e. Arizona

Similar to the preceding species, but the plant and flowers are smaller and the leaves thicker and narrower. About 13 species are known from North America.

Tailleaf Pericome

Taperleaf

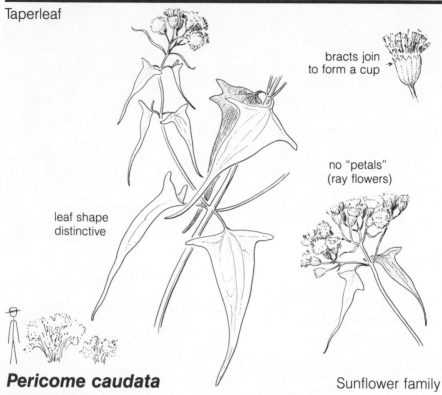

bracts join
to form a cup

no "petals"
(ray flowers)

leaf shape
distinctive

Pericome caudata   Sunflower family

Blooms yellow to orange-yellow. July-October. 6,000' - 9,000'
S. Colorado and New Mexico to s. Nevada, Arizona and California

Although rare at the highest elevations, tailleaf pericome is often locally abundant and colorful in the ponderosa pine belt. The plants often grow in clumps up to 5 feet high and, in late summer, are covered with small yellow flower heads without rays.

Perhaps the most unusual feature of the plant are the leaves which are limp and drooping, and in shape are triangular and narrowed to a long, slender point. They are dark green and are strong-scented.

Because of the goat-like odor of this plant the Spanish people in New Mexico call it *yerba del chivato,* meaning herb of the he-goat. It is thought to be very helpful in treating rheumatism; a bath of the boiled roots is used. In some cases the rheumatic limbs are rubbed with the dry leaves of the plant.

Only two species are known, both from western North America. The plants are of little horticultural importance and are offered only by dealers in western native plants. "Taperleaf" is especially beautiful along the road from Taos to Cimarron, New Mexico, in the northern part of the state.

Orange Sneezeweed

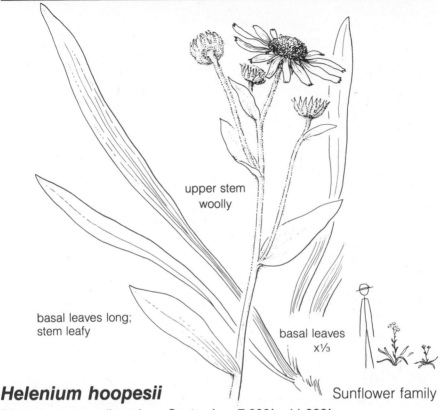

upper stem
woolly

basal leaves long;
stem leafy

basal leaves
x⅓

Helenium hoopesii

Sunflower family

Blooms orange-yellow. June-September. 7,000' - 11,000'
Wyoming to Oregon, s. to New Mexico, Arizona and California

Sometimes abundant in the rich soil of coniferous forests, the orange sneezeweed is an upright plant with large flower heads 2 inches or more in diameter.

The strong odor of the flowers causes sneezing in some people — thus the common name. The species name, *hoopesii,* is somewhat amusing, for it sounds like a sneeze itself.

A disease of sheep, called "spewing sickness," is caused by a poisonous substance called dugaldin, which is contained in the plant. Cattle are also poisoned, but rarely eat the plant.

The Navajos used the plant as a remedy for vomiting, and from the roots they make a chewing gum. A soft yellow dye is made from the flowers. In some of the more isolated parts of New Mexico the groundup roots are applied to relieve pains in the chest and shoulders due to colds or pneumonia. Stomach derangements, and even supposed cases of leprosy, are treated with preparations made from the orange sneezeweed.

Arnicas

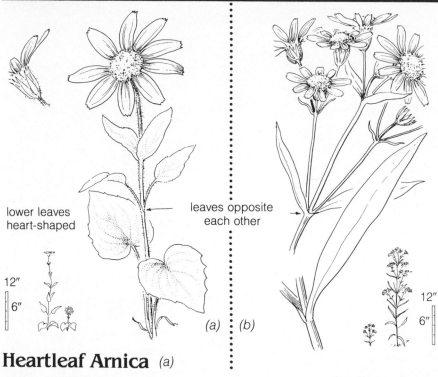

lower leaves
heart-shaped

leaves opposite
each other

12"

6"

(a) (b)

12"

6"

Heartleaf Arnica (a)

Arnica cordifolia   Sunflower family

Blooms yellow. June-August. 7,000' - 11,000'
Alaska to South Dakota, New Mexico, n. Arizona and California

A common flower in the moist pine and spruce forests of Rocky Mountain National Park in Colorado. The yellow flowers may be as much as 3 inches across, and are borne singly at the top of the stems. Large, heart-shaped leaves grow at the base of the plant but become increasingly narrower and smaller going up the stem.

The familiar tincture of arnica that is used for treating bruises and sprains is obtained from a European species of arnica.

Leafy Arnica (b)

Arnica foliosa   Sunflower family

Blooms yellow. July-August. 7,000' - 11,000'
Alaska to Colorado, New Mexico, n. Arizona and California

This species appears to be the most common one in New Mexico. It greatly resembles heartleaf arnica but the lower leaves are more elliptical in shape; there are usually three or more flowers on a stem.

Groundsel

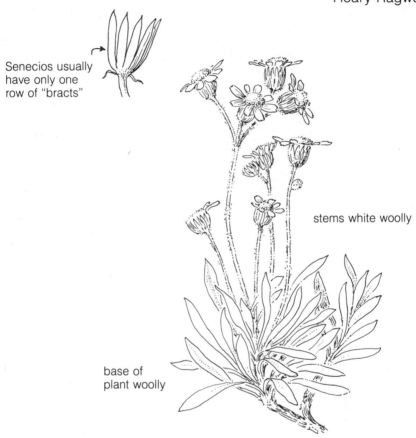

Senecios usually have only one row of "bracts"

stems white woolly

base of plant woolly

Senecio werneriaefolius

Sunflower family

Blooms yellow. June-August. 9,000' - 11,000'
South Dakota, Wyoming, Colorado, Utah and n. Arizona

The senecios comprise a large group of plants similar in appearance and difficult to identify. As a group they may be distinguished by the bracts, beneath the flower, which are in a single series only.

This species is seldom more than 1 foot high and bears many golden flowers. The plant is usually quite woolly and the cup under the flower looks as though it was covered with fur. When the plant goes to seed, long white hairs appear and it is in reference to this that the plant is called senecio from the Latin meaning "old man."

Groundsel

Bigelow Groundsel

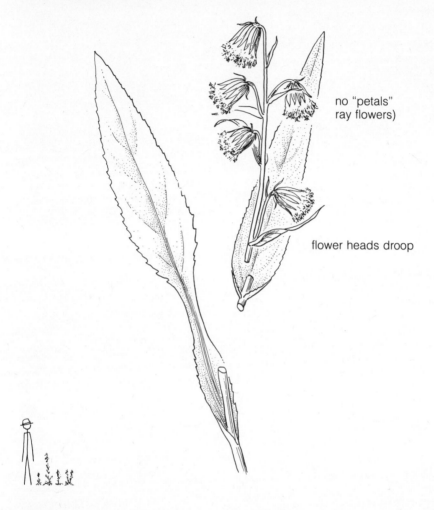

no "petals"
ray flowers)

flower heads droop

Senecio bigelovii

Sunflower family

Blooms yellow. July-September. 7,000' - 11,000'
New Mexico and Arizona

    A rather inconspicuous plant of meadows and moist forests. It bears a few drooping yellow flower heads without "petals." The bracts at the base of the head are often purplish in color. Some species are poisonous to livestock but are seldom eaten when other forage is available.

Pale Agoseris

Tall False Dandelion

12"

6"

Agoseris glauca

Sunflower family

Blooms yellow. May-October. 6,500' - 10,000'
British Columbia, s. to New Mexico, Arizona and Nevada

It is easy to confuse the false dandelions with the true dandelions, but the seeds show the difference. Those of the dandelion are rough and spiny while false dandelion seeds are smooth in comparison. The leaves of this plant are pale green and only slightly lobed. Plants reach a height of 1 to 3 feet, and often grow along brook banks and in mountain meadows. (Arizona's common variety, *laciniata,* has conspicuously-lobed leaves.)

Orange Agoseris

Burnt-Orange Dandelion

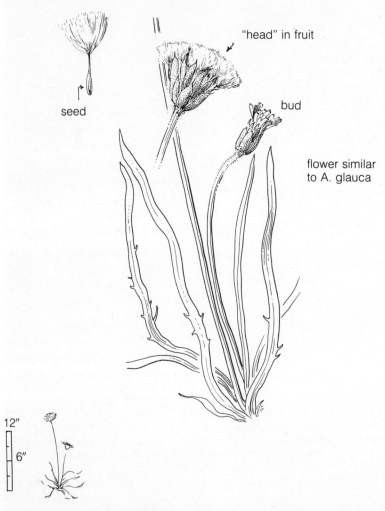

seed

"head" in fruit

bud

flower similar
to A. glauca

12"

6"

Agoseris aurantiaca   Sunflower family

Blooms orange when fresh, and drying to purple. June-August. 5,000' - 10,000'
Alberta and British Columbia to New Mexico and n. Arizona

The orange flowers and dark green leaves should distinguish orange agoseris from its close relative. A glance at the leaves shows another difference, for in this species the edges are not lobed.

Monkeyflower, *Mimulus guttatus*

Goldenpea, *Thermopsis pinetorum*

Creeping barberry, *Berberis repens*

Southwestern St. Johnswort,
Hypericum formosum

Golden columbine, *Aquilegia chrysantha*

Nelson larkspur, *Delphinium nelsonii*   Parry bellflower, *Campanula parryi*

Harebell, *Campanula rotundifolia*   Colorado columbine, *Aquilegia caerulea*

Rocky Mountain iris, *Iris missouriensis*

Rocky Mountain Iris

Western Blue Flag

12"

6"

Iris missouriensis

Iris family

Blooms blue to violet. May-September. 6,000' - 9,500'
North Dakota to British Columbia, s. to New Mexico, Arizona and California

Anyone familiar with cultivated flowers will be able to recognize the Rocky Mountain iris, for it is very similar to the common cultivated species. The plants may reach a height of 3 feet, and their large blue blossoms may literally carpet the floor of wet meadows in the Transition and Canadian Life Zones. The tall flowering stems arise from a fleshy horizontal rootstock. Due to the fact that several stems may come from one root, the plants often grow in clumps. The Navajos are reported to use a wild iris, probably this one, in making a green dye. Some Spanish New Mexicans treat smallpox with iris roots by placing a necklace of sliced iris roots around their throats.

Orris-root powder, used in toilet preparations, is obtained from one of the European species of iris. The ancient Greeks considered the iris valuable as a source of drugs, and iris preparations were prescribed for an amazing number of ailments. Now its medicinal value is limited to an extract of the roots used as an emetic and cathartic.

Colorado Columbine

Blue Columbine

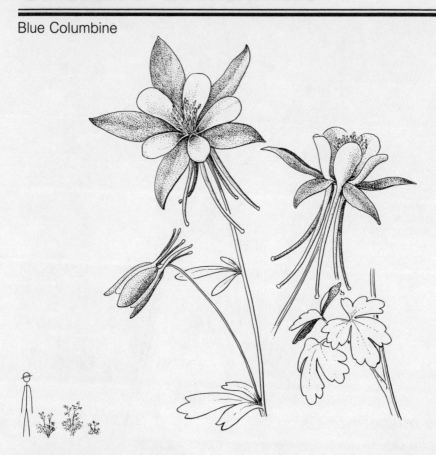

Aquilegia caerulea   Buttercup family

Blooms blue to whitish. June-July. 8,000' - 11,000'
Sw. Montana to n. New Mexico and n. Arizona

Few indeed are the flowers of the west that can compare with the Colorado columbine in beauty. The blossoms, occasionally as much as 6 inches in diameter, vary somewhat in color from blue to lavender and even white.

In typical specimens the conspicuous outer part of the flower is blue while the smaller petals arranged around the center are white.

This is the state flower of Colorado. In many places it has become a victim of its own beauty through digging and picking by thoughtless people — such vandalism is now prohibited by Colorado state law.

The flower spurs contain nectar which can be reached comfortably by only the "long-tongued" insects and hummingbirds. Some of the other insects, however, manage to get the sweet nectar by cutting holes in the spurs.

Larkspurs

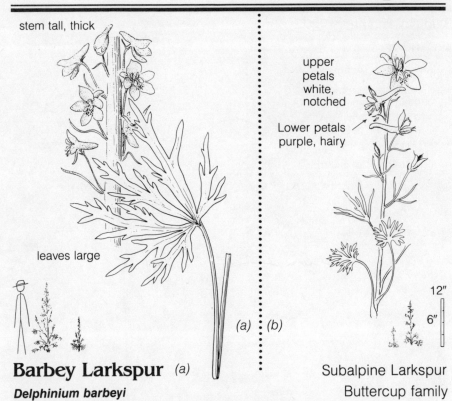

stem tall, thick

leaves large

upper petals white, notched

Lower petals purple, hairy

12"

6"

(a) : *(b)*

Barbey Larkspur *(a)*

Delphinium barbeyi

Blooms deep blue. July-August. 7,000' - 12,000'
Wyoming, Colorado, Utah, New Mexico and e. Arizona

Grows in moist rich soil of the spruce-fir belt on up to timberline and is the most handsome species of larkspur in Rocky Mountain National Park.

Larkspurs have received their name from the prolonged upper seal which resembles the spur of a bird. Spanish Californians call larkspur *"espuela del caballero,"* meaning cavalier's spur.

Subalpine Larkspur
Buttercup family

Nelson Larkspur *(b)*

Delphinium nelsonii   Buttercup family

Blooms dark purplish-blue. June-July. 6,000' - 9,000'
South Dakota to Idaho, s. to Colorado, n. Arizona and Nevada

An early blooming larkspur about 1 foot high with circular leaves deeply cut into several segments. It is a plant of the pine forests and is the common form on the Kaibab Plateau of northern Arizona.

As in other species, this one contains a poisonous substance called delphinine which may be deadly to cattle.

Columbia Monkshood

Friar's-cap, Blue-weed

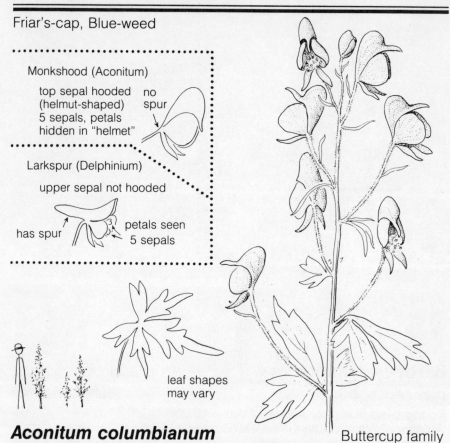

Monkshood (Aconitum)

top sepal hooded
(helmut-shaped)
5 sepals, petals
hidden in "helmet"

no
spur

Larkspur (Delphinium)

upper sepal not hooded

has spur

petals seen
5 sepals

leaf shapes
may vary

Aconitum columbianum

Buttercup family

Blooms blue. June-September. 5,000' - 10,000'
British Columbia to Montana, s. to New Mexico, Arizona and California

The beautiful dark-blue to violet flowers of the monkshood are often confused with those of the larkspur, but a close look will show you that the upper sepal is helmet-shaped in monkshood, and long and spurlike in the larkspur. The two plants are often found growing together, so it is important to be able to tell them apart.

Under favorable conditions the plants may reach a height of 6 to 7 feet, but this is unusual. Be on the lookout for this beauty along streams, in mountain meadows, or open woods.

A powerful heart stimulant called aconite is obtained from the European monkshood. All of the species of monkshood contain poisonous substances, and they may produce fatal results among stock when other feed is scarce. The plants are said to be most toxic just before flowering. The thick turnip shaped root is used medicinally and is virulently poisonous.

Lupines

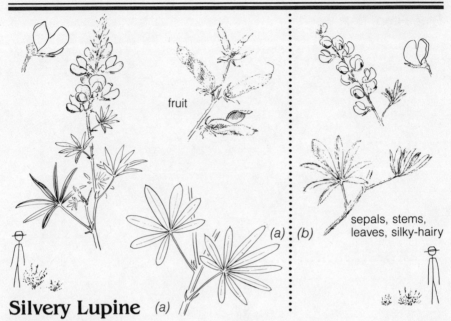

fruit

(a) (b)

sepals, stems,
leaves, silky-hairy

Silvery Lupine (a)

Lupinus argenteus   Pea family

Blooms blue to purple. June-October. 7,000' - 10,000'
North Dakota and Montana, s. to New Mexico and n. Arizona

Lupines are among the best known flowers in the Southwest. Most of the lupines are found at lower elevations, but the silvery lupine is at home in the rich soil of open ponderosa pine forests.

Some of the species are poisonous to livestock, particularly sheep. The toxic agent seems to be concentrated in the seeds and pods.

In Texas the lupines are called bluebonnets. About 200 species are found in this country, most of them growing in the west. Some species are used in horticulture because of their ornamental value. The seeds of these attractive plants are eaten by several species of western upland game birds.

Palmer Lupine (b)

Lupinus palmeri   Pea family

Blooms violet to blue. April-October. 4,000' - 8,000'
New Mexico and Arizona

In Arizona this is the most common and widespread of the lupines in ponderosa pine forests of the Transition Life Zone.

Lupines may be distinguished by the pea-like flowers, the leaves consisting of several leaflets radiating from the center, and the fact that they are never trailing, tendril-bearing, or twining.

Lewis Flax

Blue Flax

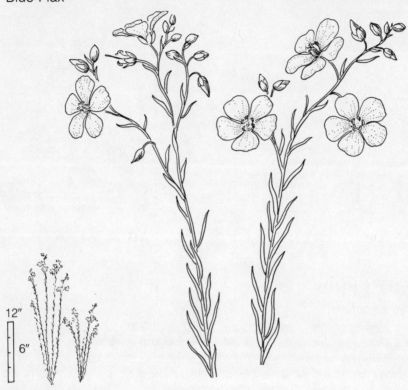

Linum lewisii

Flax family

Blooms sky blue. March-September. 3,500′ - 9,500′
Saskatchewan and Alaska to n. Mexico

A slender plant from 1 to 2 feet tall with several erect stems and small, narrow leaves. The beautiful flowers, which are normally sky blue, may sometimes be almost white. They are about 1 inch across, and have veins of darker blue.

From very early times cultivated flax, *Linum usitatissimum,* has furnished the world with linen from its fiber and linseed oil from its seeds. Although of no commercial value, our native Lewis flax has found several uses.

The Klamath Indians cultivate the plant for its strong fibers used in baskets, mats, fish nets, etc. In New Mexico some of the old-timers of Spanish descent use the flax seeds as a poultice to treat infected wounds and reduce swelling and boils. The poultice is made by grinding dry flax seeds, mixing one teaspoonful of corn meal, and adding enough boiling water to make a paste.

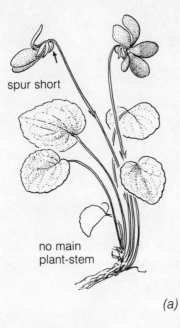

tiny "leaflets"
on flower stems
narrow, sharp, pointed

spur short

no main
plant-stem

(a)

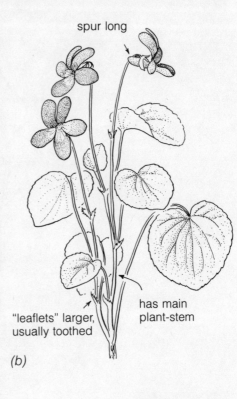

spur long

"leaflets" larger,
usually toothed

has main
plant-stem

(b)

Wanderer Violet *(a)*

Viola nephrophylla

Blooms violet. April-June. 5,000' - 9,500'
Canada to New Mexico, Arizona and California

Meadow Violet
Violet family

The wanderer violet is most often found in moist meadows, but also in canyons and on mountain slopes up to the spruce belt. Often reddish underneath, its leaves arise from the base of the plant and are rather thickish.

Hook Violet *(b)*

Viola adunca

Blooms blue. June-July. 7,000' - 9,500'
Canada to New Mexico, Arizona and California

Blue Violet
Violet family

Widely distributed in damp mountain forests, the hook violet is a small plant only a few inches high. The flowers are usually blue but are sometimes white and commonly measure less than one-half inch across.

Parry Gentian

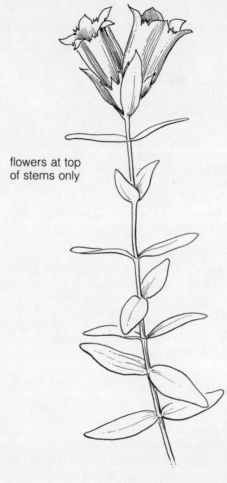

flowers at top
of stems only

12″

6″

Gentiana parryi

Gentian family

Blooms deep blue. August-September. 8,500′ - 12,000′
Wyoming and Utah to New Mexico and e. Arizona

Upright flower stems about 1 foot high bearing from one to five flowers, blue when open and blackish when closed. The flowers are often partly enveloped by rather broad leaves. It is such a sun-loving plant that the flowers close when the sun goes under a cloud.

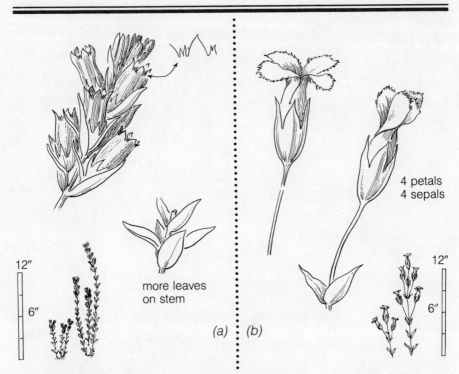

4 petals
4 sepals

12″
6″

more leaves
on stem

12″
6″

(a) : *(b)*

Rocky Mtn. Pleated Gentian *(a)*

Gentiana affinis

Marsh Gentian
Gentian family

Blooms deep blue or violet. August-October. 7,000′ - 9,500′
Saskatchewan to British Columbia, s. to Colorado and California

The Navajos are said to use a preparation of this plant for treating head-aches. The flowers are narrow and funnel-shaped, and are about 1 inch long. They are borne along the upper one-third of the flower stalk.

Rocky Mtn. Fringed Gentian *(b)*

Gentiana thermalis

Gentian family

Blooms deep blue. August-September. 8,000′ - 11,000′
W. Canada to New Mexico and Arizona

One of the most beautiful of all mountain flowers, the Rocky Mountain fringed gentian is often found in moist, sunny meadows. Its flower stalks reach a height of about 1 foot, and each bears a handsome flower 1 to 2 inches long with four fringed petals. The buds are twisted clockwise.

Gentian root has been used medicinally for centuries as a mild stimulant of gastric secretions.

Manyflowered Gilia

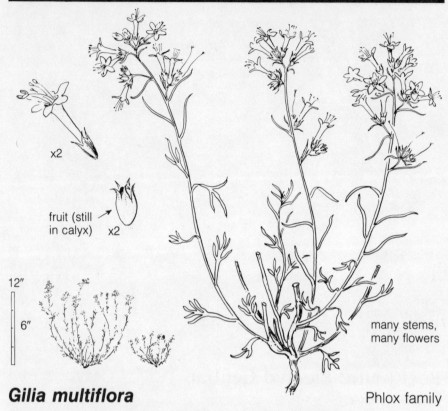

x2

fruit (still
in calyx) x2

12"

6"

many stems,
many flowers

Gilia multiflora

Phlox family

Blooms pale violet to blue. July-October. 4,000' - 9,000'
New Mexico to s. Nevada and Arizona

A low, branching, somewhat woody plant having small bluish flowers less than one-inch across. The rough-textured stems have many rather small, narrow, pointed leaves which are dull green in color. The conspicuous stamens projecting beyond the petals give the flowers a light feathery appearance.

There is often confusion regarding the correct pronunciation of the generic name of this plant. Although common usage makes acceptable such pronunciations as GILL-ee-uh or JILL-ee-uh, actually the name should be pronounced HEEL-ee-uh. The genus was named after Felipe Luis Gil, a Spanish botanist, and thus a Spanish pronunciation would be correct.

Practically all members of the phlox family are native to America, and many are beautiful enough to warrant cultivation as ornamentals.

About 120 species of gilia are recognized in North America, the majority in the United States, and these mainly in the west. The seeds and foliage are utilized to a small extent by upland game birds and some rodents.

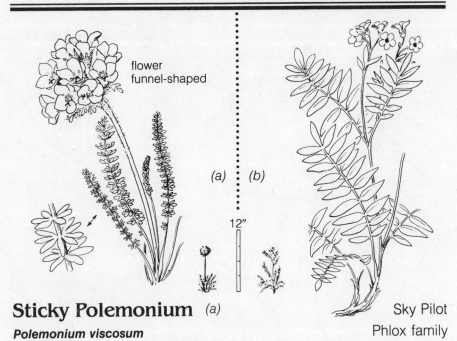

flower
funnel-shaped

(a) : (b)

12"

Sticky Polemonium (a)

Sky Pilot
Phlox family

Polemonium viscosum

Blooms blue to purplish. June-September. 9,000' - 13,500'
Wyoming to Washington and n. Arizona

A plant having stems seldom more than 10 inches high, and bearing heads of funnel-shaped or tubular blue flowers with bright orange anthers. This plant, sometimes called sky pilot, has been found growing at elevations higher than 13,000 feet on some of the high mountains of Colorado. Some species are cultivated in flower gardens, and it is said that all polemoniums are easily raised from fall-sown seed. They may also be propagated by division.

Skunkleaf Polemonium (b)

Jacob's Ladder
Phlox family

Polemonium delicatum

Blooms violet-blue. June-August. 9,000' - 12,000'
Idaho to New Mexico and Arizona

Sometimes found growing with the preceding species, this plant may be distinguished from it by the white or yellow throat of the flower. The blue petals are also spread out more widely, and the flower, as a whole, is more bell-shaped than tubular. The leaves are sometimes mistaken for fern leaves and from the ladder-like arrangement of the leaflets, it is easy to see why the plant is sometimes called Jacob's ladder. As in most polemoniums, the plant has a skunk-like odor.

Bluebells

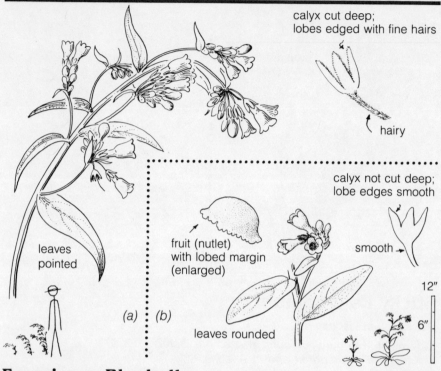

calyx cut deep; lobes edged with fine hairs

hairy

leaves pointed

(a) *(b)*

calyx not cut deep; lobe edges smooth

fruit (nutlet) with lobed margin (enlarged)

smooth

leaves rounded

12″

6″

Franciscan Bluebells *(a)*

Mertensia franciscana

Lungwort
Borage family

Blooms blue. June-September. 7,000′ - 12,000′
W. Colorado and se. Utah to New Mexico and Arizona

The various species of *Mertensia* are easily recognized by their many pendent, tubular flowers. This plant grows in clusters, and the stems bearing flowers are usually about 1 foot high. Before maturity the flowers may be pinkish to white in color. This species has pointed leaves with short, flattened hairs on the upper surface. Look for this attractive plant along streams or in other moist places associated with pine, spruce, or aspen.

MacDougall's Bluebells *(b)*

Mertensia macdougallii

Borage family

Blooms blue. March-June. 6,000′ - 9,000′
Arizona

Although restricted in distribution to the state of Arizona, this plant is often locally quite abundant. It should be looked for in the ponderosa pine belt of Arizona, and it is recorded on the south rim of Grand Canyon.

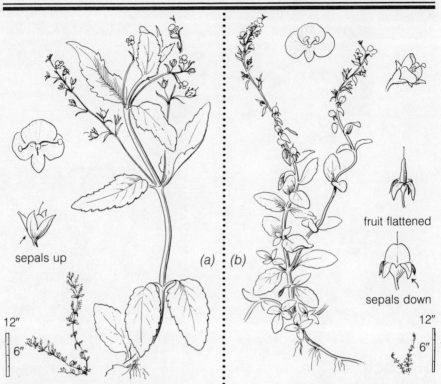

sepals up

(a) (b)

fruit flattened

sepals down

12″

6″

12″

6″

American Speedwell (a)

American Brooklime
Snapdragon family

Veronica americana

Blooms pale blue. May-August. 1,500′ - 10,000′
Throughout most of North America

Several kinds of *Veronica* are found in the Southwest, all rather low herbs having blue, white or pinkish flowers. After seeing these small plants, it is hard to imagine some of the tropical relatives which are of tree size. The genus was named in honor of Saint Veronica.

American speedwell is semi-aquatic and will be found in and around springs or streams.

Speedwell (b)

Alpine Speedwell
Snapdragon family

Veronica wormskjoldii

Blooms sky blue. July-September. 9,500′ - 12,000′
Greenland to Alaska, s. to New Hampshire, New Mexico and Arizona

Distinguished from the preceding species by darker flowers at the top of the stem. It is not aquatic, but grows in damp spots on high mountains.

Bellflowers

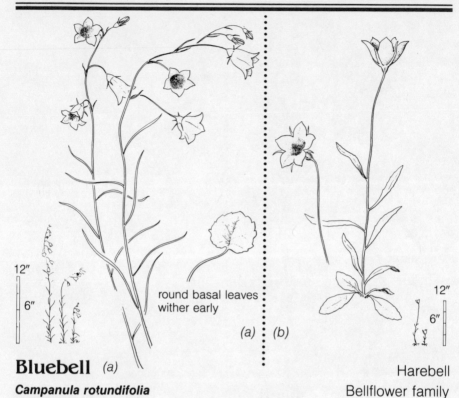

12″
6″

round basal leaves
wither early

(a) : (b)

12″
6″

Bluebell *(a)*

Harebell

Campanula rotundifolia

Bellflower family

Blooms deep blue. June–September. 8,000′ - 12,000′
N. part of America and Europe, s. in mountains to Arizona and California

With their drooping, bell-shaped flowers the bluebells are among the most beautiful of mountain flowers. The slender flower stem springs from a cluster of roundish leaves which usually wither before the flowers bloom.

This same species grows in Scotland and is the well-known bluebells of Scotland.

Parry Bellflower *(b)*

Purple Bellflower

Campanula parryi

Bellflower family

Blooms purplish-blue. July–September. 7,000′ - 10,000′
Wyoming and Utah to n. New Mexico and n. Arizona

It is said that pregnant Navajo women eat this plant to insure the birth of a girl. The Zuñis chew the blossoms and then apply the material to the skin to remove excess hair. It is not likely, however, that this method of depilation will supplant the present use of razors. The chewed root is also used in treating bruises.

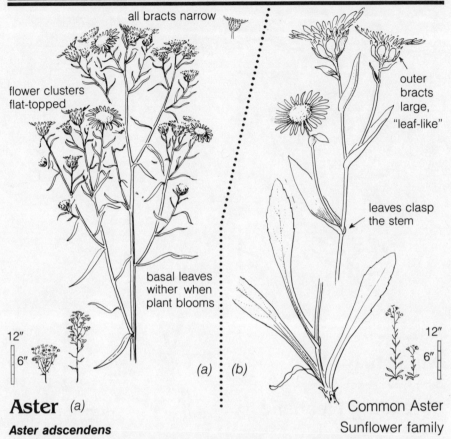

all bracts narrow

flower clusters
flat-topped

outer
bracts
large,
"leaf-like"

leaves clasp
the stem

basal leaves
wither when
plant blooms

12"
6"

(a) (b)

12"
6"

Aster *(a)*

Aster adscendens

Common Aster

Sunflower family

Blooms blue, yellow centered. August-September. 6,500′ - 8,500′
Saskatchewan to Washington, s. to Colorado and n. Arizona

A low plant with narrow leaves usually less than one-half inch wide. It
grows in mountain meadows and openings in coniferous forests.

Leafybract Aster *(b)*

Aster foliaceus

Sunflower family

Blooms purple with yellow center. August-September, 7,500′ - 10,000′
Wyoming to British Columbia, s. to New Mexico and central Arizona

Asters are often confused with fleabane daisys; however, asters have
several rows of bracts beneath the flower heads while the fleabanes usually
have only one or two rows. The ray flowers or "petals" of the fleabanes are
narrower and more numerous than those of asters. These differences are
shown in the small drawing on page 117.

Fleabanes

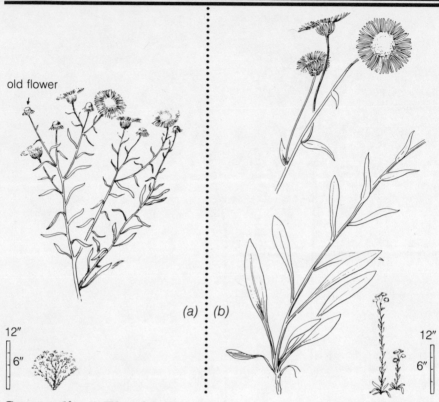

old flower

(a) : (b)

12"
6"

12"
6"

Spreading Fleabane (a)
Erigeron divergens

Branching Daisy
Sunflower family

Blooms bluish, yellow centered. February-October. 1,000' - 9,000'
British Columbia s. to Texas, New Mexico, Arizona and California

Probably the most common fleabane daisy at the lower elevations of
the Southwest mountains. It is seldom over a foot high and bears many flowers, each having 100 or more rays or "petals."

Aspen Fleabane (b)
Erigeron macranthus

Blue Aspen Daisy
Sunflower family

Blooms dark blue to pale violet, yellow centered. July-October. 6,000' - 9,500'
W. Canada s. to New Mexico and Arizona

A coarse plant reaching a height of 3 feet under favorable conditions.
It may be of interest to know that the word "daisy" is a corruption of
day's eye, which refers to the resemblance of the flowers with their yellow
centers and rays to the sun and its rays.

Alpine Daisy

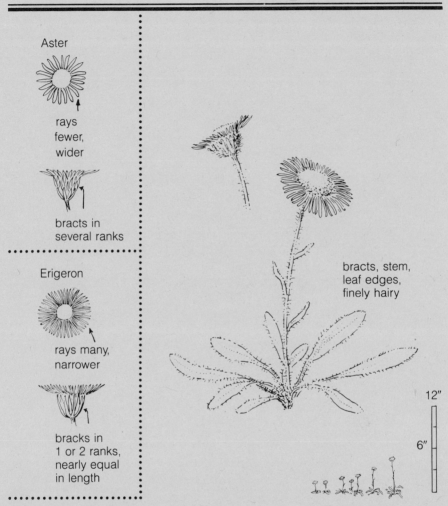

Aster

rays
fewer,
wider

bracts in
several ranks

Erigeron

rays many,
narrower

bracks in
1 or 2 ranks,
nearly equal
in length

bracts, stem,
leaf edges,
finely hairy

12″

6″

Erigeron simplex   Sunflower family

Blooms violet with yellow centers. August-September. 10,000′ - 13,000′
Montana to n. New Mexico, n. Arizona and California

    This is a daisy of the Alpine Life Zone. As is true of many alpine plants, it has stems only a few inches high, and its growth is somewhat mat-like. Occasional flowers are found that have very small rays, but this characteristic is not constant. The alpine daisy seems to be most common around high mountain lakes.

117

Pacific Anemone

Windflower

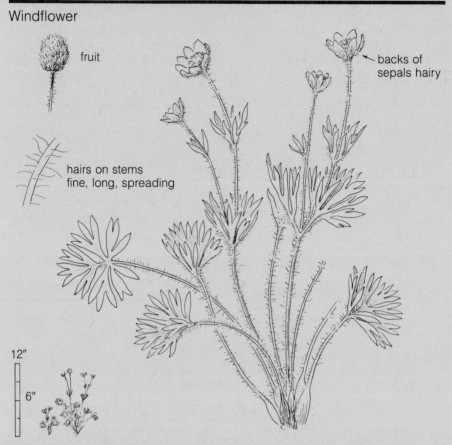

fruit

backs of
sepals hairy

hairs on stems
fine, long, spreading

12"

6"

Anemone globosa

Buttercup family

Blooms usually reddish-purple. July. 10,000' - 12,000'
W. Canada to New Mexico, n. Arizona and California

Anemones are attractive plants, growing in cold and temperate regions almost everywhere. Some of the Old World species are popular garden flowers, but the native species are seldom cultivated. Pretty as they are, the flowers have no true petals, only sepals which resemble petals. Many anemones have a bitter juice which irritates the skin, and may be poisonous if taken internally.

Practically all parts of the plant are woolly and perhaps this is a blessing, for the mountain tops on which they grow are usually a bit chilly. The name anemone comes from the Greek meaning "flower shaken by the wind." Hence one common name "windflower." The flowers are usually purplish on the outside, but may be either purplish or yellowish within.

Rocky Mountain Clematis

Alpine Clematis

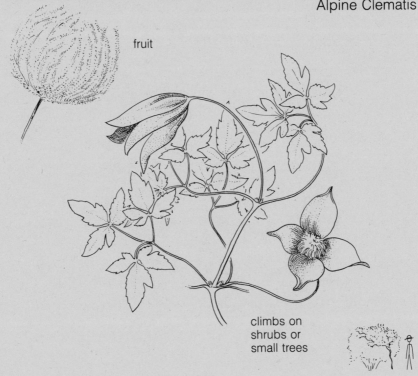

fruit

climbs on
shrubs or
small trees

Clematis pseudoalpina

Buttercup family

Blooms purple or violet. July. 7,000' - 9,000'
South Dakota, and Montana to New Mexico and Arizona

With its four large, purple, petal-like sepals this is one of the most attractive of the many different species of clematis in the Southwest. The alpine clematis may be found growing in the rich soil of coniferous forests in the Transition and Canadian Life Zones. It is usually low and trailing, but under favorable conditions it may clamber, vine-like, over low shrubs and trees.

It is a handsome plant even after the flowers have gone, for the seeds have long silky plumes which form feathery clusters that are conspicuous and ornamental. Some species are grown as ornamentals, and some of the exotic species are among the most beautiful of cultivated climbing plants.

About 150 species of clematis are known; they have a wide geographical distribution, although they are most abundant in temperate regions.

Another species with smaller, cream-colored flowers, *Clematis ligusticifolia,* was formerly used by Indians as a remedy for colds and sore throats.

119

American Vetch

Purple Vetch

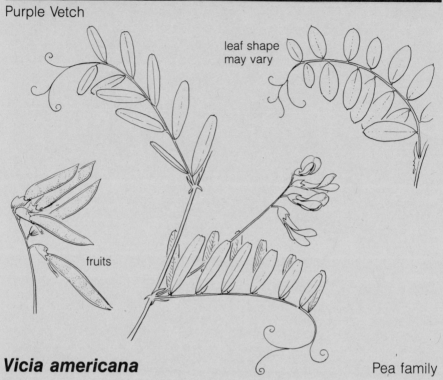

leaf shape
may vary

fruits

Vicia americana   Pea family

Blooms purplish to blue. May-September. 5,000' - 10,000'
Canada to Virginia, New Mexico, Arizona and California

Slender scrambling stems from 1 to 3 feet long, purple, pea-like flowers, and compound leaves ending in tendrils serve to identify American vetch.

Some of the Old World species of vetch are cultivated in the United States for hay and as cover crops. The native species, however, are soon killed out by close grazing. This pretty vine-like plant is usually quite common in pine forests throughout the Southwest.

Both the seeds and foliage are eaten to a limited extent by birds and rodents. Leaves of American vetch and other species make up about 5% of the summer diet of the dusky blue grouse in the Rocky Mountain area. Other birds making slight use of the plant include mourning doves, ring-necked pheasants, prairie chickens, quail and wild turkeys.

The Indians of California and New Mexico are reported to have baked or cooked the young stems of this plant for greens.

About 150 species of *Vicia* are widely distributed in the Northern Hemisphere and some in South America. There are about 24 species in North America, of which some are introduced. The species are mostly cool season plants and are quite easy to cultivate.

New Mexico Locust

Rose Locust

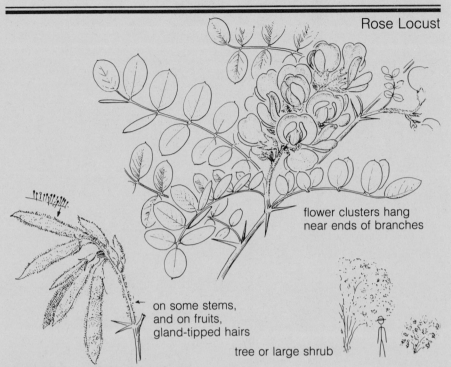

flower clusters hang
near ends of branches

on some stems,
and on fruits,
gland-tipped hairs

tree or large shrub

Robinia neomexicana

Pea family

Blooms purplish-pink. May-July. 4,000' - 8,500'
S. Colorado to s. Nevada, w. Texas, New Mexico, Arizona and n. Mexico

Throughout the southwestern mountains this plant is quite common, but it is especially abundant and colorful on the Kaibab Plateau of northern Arizona. The large and showy flowers grow in clusters at the ends of slender branches. It is the most handsome species of locust, and is well worthy of cultivation as an ornamental.

Occasional plants may reach a height of 25 feet, although typical specimens are much smaller and would be classed as large shrubs. Because of its habit of sprouting from roots or stumps, and forming thickets, it is valuable in controlling erosion.

Cattle are reported to relish the flowers, and both livestock and deer find the foliage very palatable. However, the bark, roots, and seeds are said to be poisonous. Durable fence posts may be made from the trunks.

New Mexico locust is used by the Hopis as an emetic and for treating rheumatism, and another source states that the flowers were eaten with no preparation by some of the Indians in New Mexico.

The genus *Robinia* was named after Jean and Vespasien Robin, herbalists to the King of France in the sixteenth and seventeenth centuries.

New Mexican Checkermallow

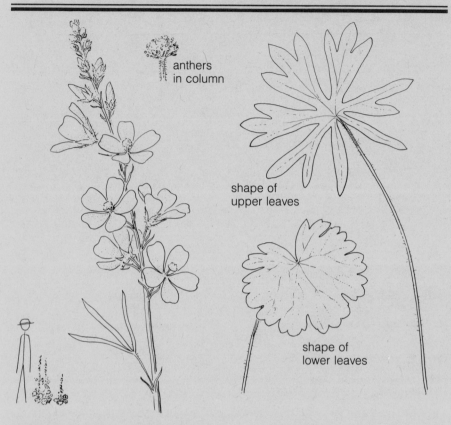

anthers in column

shape of upper leaves

shape of lower leaves

Sidalcea neomexicana

Mallow family

Blooms purplish-pink to mauve. June-September. 5,000' - 9,500'
Wyoming and Idaho to n. Mexico and California

Although not common, the New Mexican checkermallow may occasionally be found growing in wet meadows and along streams in the ponderosa pine belt. It is an attractive plant, reaching a height of 3 feet.

Both flowers and seeds bear a marked resemblance to those of cultivated hollyhocks, which also belong to the mallow family.

The smooth dark-green leaves on the upper part of the plant are deeply cut into several segments which radiate from the center. The lower leaves, however, are almost round, or have only rounded teeth on their edges.

The mallow family contains such economically important plants as cotton and okra and some species with showy flowers which are cultivated as ornamentals. A European species of mallow, the marshmallow *(Althaea)* was originally used in making the well-known confection.

Indians are said to use the checkermallow for greens.

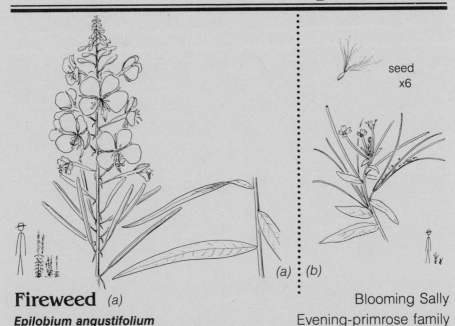

seed
x6

(a) : *(b)*

Fireweed *(a)*

Blooming Sally

Epilobium angustifolium

Evening-primrose family

Blooms rose to lilac. July-September. 7,000' - 11,500'
Widely distributed in the Northern Hemisphere

Large areas in the high mountains are often ablaze with the blossoms of fireweed. The plant is from 1 to 5 feet high and, as the common name indicates, is often found growing with aspens on burned-over areas, although it also occurs in open woods and meadows. Seeds are borne in long, podlike capsules, and each seed bears a tuft of hairs. Thus the wind easily picks up the seeds and carries them into areas that have been devastated by forest fires.

The leaves are narrow and long, and the veins make a scalloped pattern near the smooth edges of the leaf. This species may be distinguished from most of the others by the separation of the petals clear to their bases at the tip of the seed capsule.

Sticky Willowweed *(b)*

Tall Cottonweed

Epilobium adenocaulon

Evening-primrose family

Blooms purplish-pink, or sometimes almost white. June-September. 4,000' - 9,500'
E. United States to British Columbia, Arizona and Mexico

Resembles fireweed, but has smaller flowers with the petals united at base into a tube. Silky, hairy seeds give this plant one common name, "cottonweed."

Primroses

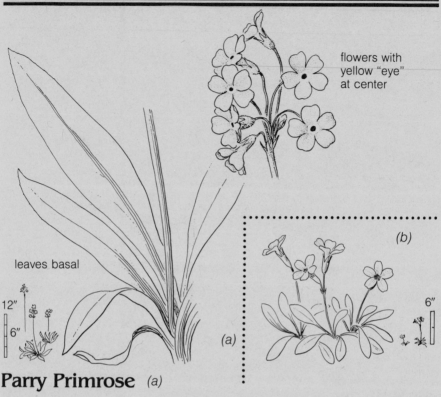

flowers with yellow "eye" at center

leaves basal

12"

6"

(a)

(b)

6"

Parry Primrose (a)

Primula parryi   Primrose family

Blooms reddish-purple. June-August. 10,000' - 12,000'
Montana and Idaho to New Mexico and Arizona

One of the most striking plants of the high mountains, growing from 1 to 2 feet tall and bearing clusters of brilliant flowers. The smooth, thick leaves, which are quite long, all grow in a rosette at the base of the plant.

Look for the Parry primrose among rocks in the Hudsonian and Alpine Zones and pay particular attention to brooks and other moist places. You'll be in for a disappointment if you think the flowers will smell as good as they look, for they have the distinct odor of carrion.

Colorado Primrose (b)   Fairy Primrose

Primula angustifolia   Primrose family

Blooms rose purple with a yellow center. July. 10,000' - 13,000'
Colorado and n. New Mexico

A dwarf, alpine species only a few inches high. This and other alpine plants bloom in midsummer — which is really spring at such high elevations.

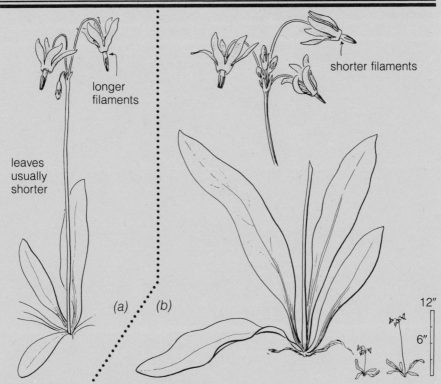

longer filaments

shorter filaments

leaves usually shorter

(a) (b)

12"

6"

Darkthroat Shootingstar *(a)*

Dodecatheon pauciflorum   Primrose family

Blooms reddish purple to pale pink. June-August. 7,000' - 11,000'
W. Canada s. to Colorado and California

    To be looked for in mountain meadows and in moist places at lower elevations. The plants are from 6 to 12 inches high and bear a loose cluster of flowers at the end of a reddish, leafless stem.

    Leaves are dull green, thickish, often much more than twice as long as wide, and are found at the base of the plant. The dart-like, yellow-centered flowers with their five turned back petals give the plant its common name.

Southern Shootingstar *(b)*

Dodecatheon pulchellum   Primrose family

Blooms reddish purple. June-August. 6,000' - 10,000'
South Dakota and Wyoming to New Mexico and Arizona

    Very similar to the preceding species and separated from it largely on the basis of longer leaves and shorter filaments in the center of the flower.

Penstemons

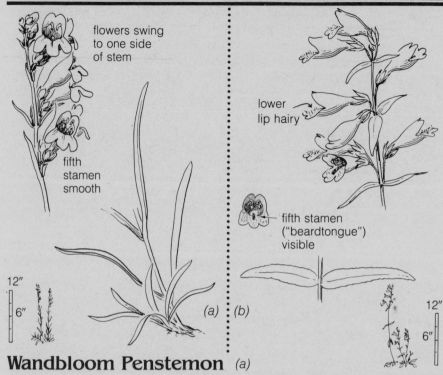

flowers swing to one side of stem

fifth stamen smooth

lower lip hairy

fifth stamen ("beardtongue") visible

12"
6"

12"
6"

(a) *(b)*

Wandbloom Penstemon *(a)*

Penstemon virgatus   Snapdragon family

Blooms usually pale violet with deep purple lines. Blooming throughout the summer. 5,000' - 10,000'
New Mexico and Arizona

   A common and very handsome penstemon growing in the pine woods and mountain meadows of Arizona and New Mexico. The typical plant has long, narrow leaves; the one tongue-like stamen lacks the hairs found in the other species.

Whipple Penstemon *(b)*   Dusky Beardtongue

Penstemon whippleanus   Snapdragon family

Blooms dark purple or wine colored, sometimes yellow or white. July-August. 6,500' - 12,000'
S. Montana to New Mexico and n. Arizona

   Ordinarily the plants reach a height of about 2 feet, but at very high altitudes they may be dwarfed. Many of the western species of *Penstemon* are showy in flower and are worthy of cultivation. Other than this they have little value, although some are browsed by animals.

Elephanthead Pedicularis

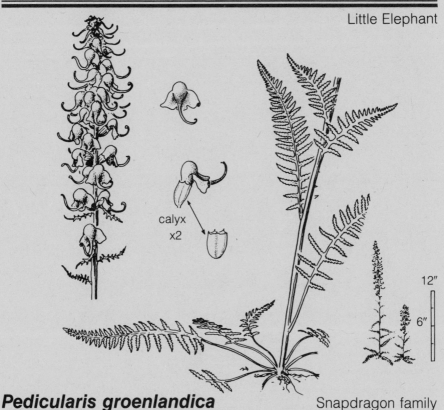

calyx
x2

12"

6"

Pedicularis groenlandica

Snapdragon family

Blooms pinkish purple. July-August. 8,000' - 12,000'
Greenland to Alaska, s. in mountains to New Mexico, e. Arizona and California

Growing in marshy mountain meadows, elephanthead pedicularis is one of the most interesting flowers in the Southwest mountains. A glance at the flower will explain its common name: the upper petal forms a long, down-curved tube having a comical resemblance to an elephant's trunk.

The flowers, which are about one-half inch long, are crowded on upright stems from 6 inches to 2 feet in height.

Many plants cultivated for their ornamental flowers are members of the snapdragon family, but few are of any other economic value. However, the foxglove, which is native to the Old World, is the source of the drug, digitalis, a heart stimulant.

About 250 species of *Pedicularis* grow in many parts of the Northern Hemisphere, many of them in arctic and alpine regions. A few species are found in South America and the genus is well-represented in Asia. The United States can claim about 40 native species.

127

Thistles

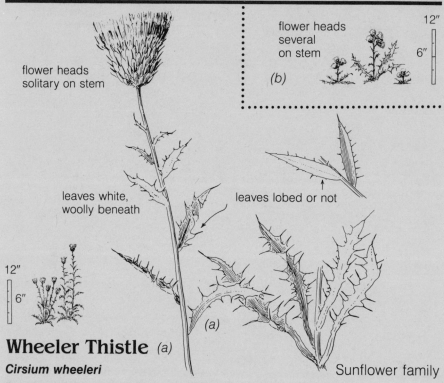

flower heads
solitary on stem

flower heads
several
on stem

12″

6″

(b)

leaves white,
woolly beneath

leaves lobed or not

12″

6″

(a)

Wheeler Thistle *(a)*

Cirsium wheeleri   Sunflower family

Blooms purple or lavender. June-October. 5,000′ - 9,000′
S. New Mexico and Arizona

Nearly everyone is familliar with the thistles, which are easily recognized by their spiny leaves and typical flower heads. In this species the heads usually occur singly at the ends of stems and branches. The bracts below the flower are quite broad and oval in shape.

Wheeler thistle is often quite common in open pine forests.

Drummond Thistle *(b)*

Cirsium drummondii   Sunflower family

Blooms usually straw-colored, but quite often tinged with purple. July-September. 7,000′ - 9,000′
Saskatchewan to British Columbia, s. to Arizona and California

The flower heads in this species are normally crowded in clumps on the end of the stems. Their color may vary from white to purplish. The Indians of Wyoming, Montana, Utah and Nevada are reported to have eaten the roots and stalks of Drummond thistle. The Navajos and Hopis use medicinal preparations made from thistles for treating various disorders.

Stephen Trimble

Aster, *Aster adscendens*

Wheeler thistle, *Cirsium wheeleri*

Supplementary Reading

ARMSTRONG, MARGARET
　　1915. *Field Book of Western Wildflowers*. New York: G.P. Putnam's Sons.
ASHTON, RUTH E.
　　1933. *Plants of Rocky Mountain National Park*. Washington, D.C.: United Government Printing Office.
BAILEY, H.E. and BAILEY, V.L.
　　1941. *Forests and Trees of the Western National Parks*. Washington, D.C.: United States Government Printing Office.
CLEMENTS, EDITH S.
　　1926. *Flowers of Mountain and Plain*. New York: H.W. Wilson Company.
COLLINGWOOD, G.H. and BRUSH, WARREN D.
　　1947. *Knowing Your Trees*. American Forestry Association.
ELMORE, FRANCIS H.
　　1976. *Shrubs and Trees of the Southwest Uplands*. Globe: Southwest Parks and Monuments Association.
HARRINGTON, H.D.
　　1954. *Manual of the Plants of Colorado*. Denver: Sage Books.
KEARNEY, T.H. and PEEBLES, R.H., Supplement by HOWELL, J.T. and McLINTOCK, E.
　　1960. *Arizona Flora*. Berkeley: University of California Press.
LITTLE, ELBERT L., Jr.
　　1979. *Checklist of United States Trees — Native and Naturalized*. Agricultural Handbook 541. Washington, D.C.: U.S. Department of Agriculture.
LITTLE, ELBERT L., Jr.
　　1980. *The Audubon Society Field Guide to North American Trees, Western Region*. New York: Alfred A. Knopf.
MARTIN, A.C., SIM, H.S. and NELSON, A.L.
　　1951. *American Wildlife and Plants,* New York: McGraw-Hill Book Co.
MARTIN, W.C. and HUTCHINS, C.R.
　　1980. *A Flora of New Mexico*. 2 vol. Germany: J. Cramer.
McDOUGALL, W.B.
　　1947. *Plants of Grand Canyon National Park*. Bulletin 10. Grand Canyon: Grand Canyon Natural History Association.
McDOUGALL, W.B.
　　1973. *Seed Plants of Northern Arizona*. Flagstaff: Museum of Northern Arizona.
McDOUGALL, W.B. and BAGGLEY, HERMA A.
　　1936. *Plants of Yellowstone National Park*. Washington, D.C.: United States Government Printing Office.
MUNZ, P.A. and KECK, D.D.
　　1959. *A California Flora*. Berkeley: University of California Press.
PESMAN, M. WALTER
　　1946. *Meet the Natives*. 3rd ed. Denver: Smith Brooks Printing Company.

PHILLIPS, ARTHUR M.
    1979. *Grand Canyon Wildflowers.* Grand Canyon: Grand Canyon Natural History Association.
PRESTON, R.J.
    1947. *Rocky Mountain Trees.* Ames: Iowa State College Press.
RICKETT, H.W.
    1970. *Wildflowers of the United States, vol. 4, The Southwestern States.* New York: McGraw Hill.
RICKET, H.W.
    n.d. *Wildflowers of the United States, vol. 6, The Rocky Mountain Region.* New York: McGraw Hill.
STANDLEY, P.C.
    1915. *Plants of Glacier National Park.* Washington, D.C.: United States Government Printing Office.
SUDWORTH, GEORGE B.
    1908. *Forest Trees of the Pacific Slope.* Washington, D.C.: United States Government Printing Office.

Index

NOTE: **Accepted common names** and **page references** to the color illustrations are in **bold type;** *Latin names* are in ***italicized bold type;*** *other Latin names* are in *italics;* and other common names are in regular type.

134

135

137

139